T0252737

Disturbance Ecology and Biological Diversity

Front cover image captions

Top Row:

Left: An American pika (*Ochotona princeps*) serves as an illustrative and iconic indicator of how contemporary global change can influence species distributions. The individual pictured is at a local high point in Lamoille Canyon, Ruby Mountains, northeastern Nevada. (Photo by Shana Weber, Princeton University. With permission.)

Center: Kudzu (*Pueraria montana*) provides a graphic illustration of to what degree invasive species can radically alter the composition, structure, and function of ecosystems. Pictured, its foliage is totally encompassing the canopy of the native vegetation.

Right: Domestic cattle grazing in Mojave National Preserve, southeastern California. Given that temperatures can exceed 50 °C in this region, cows spend most time at the artificial stock tank to obtain water; usage declines rapidly, with increasing distance from the well. Grazing disturbance is discussed in Chapter 4, which includes the Mojave case study where responses to cattle removal were investigated. (Image by Erik Beever.)

Middle Row:

Left: The edge of a glacier on Mt. Kilimanjaro (known locally as Uhuru Peak), Tanzania, Africa. Tropical glaciers are projected under contemporary global change to be gone altogether in very few decades. (Photo by Erik Beever.)

Center: Bobcats (*Lynx rufus*) are one of several ecological components used to assess ecosystem condition and function, in reclamation of coal-mined lands in the Appalachian region, eastern United States. (Photo by Suzanne Prange.)

Right: A moist forest on Africa's largest and tallest mountain, Mt. Kilimanjaro. Both the biological diversity and ecosystem functions on this mountain are fundamentally shaped by interactions of climate and land-use patterns. Tropical ecosystems are discussed in Chapter 5. (Photo by Erik Beever.)

Bottom Row:

Left: A diverse forest assemblage in Marquette, MI, northern United States. The yellow trees in the foreground are an aspen stand, and the surrounding older-growth trees occur in areas that haven't been disturbed as recently (see Chapter 2). (Photo by Alexandra Locher.)

Center: In Serengeti National Park, researchers begin trying to habituate lions (*Panthera leo*) to human presence as soon as they are cubs, which allows such pictures to be taken at distances of 12–15 m away with no telephoto lens. Some resources in protected areas are altered by disturbances created by tens to millions of visitor-days, annually. (Photo by Erik Beever.)

Right: Fire occurring in Basin sagebrush (*Artemisia tridentata*) shrubland, one of North America's ~20 most-endangered ecosystem types. Spatial extent and intensity of aridland fires has increased dramatically in recent decades, due to drier conditions (resulting from hotter temperatures) and the introduction of the invasive cheatgrass (*Bromus tectorum*), which dramatically reduces the fire-return interval and conscripts soil-based resources from native plants in the system. Wildfire disturbance is discussed in Chapter 3. (Photo by Scott Shaff.)

Disturbance Ecology and Biological Diversity

Scale, Context, and Nature

Edited by
Erik A. Beever, Suzanne Prange and
Dominick A. DellaSala

CRC Press
Taylor & Francis Group
Boca Raton London New York

CRC Press is an imprint of the
Taylor & Francis Group, an **informa** business

CRC Press
Taylor & Francis Group
6000 Broken Sound Parkway NW, Suite 300
Boca Raton, FL 33487–2742

© 2020 by Taylor & Francis Group, LLC

CRC Press is an imprint of Taylor & Francis Group, an Informa business

No claim to original U.S. Government works

Printed on acid-free paper

International Standard Book Number-13: 978-1-4822-9871-0 (Hardback)
International Standard Book Number-13: 978-0-367-86177-3 (Paperback)

This book contains information obtained from authentic and highly regarded sources. Reasonable efforts have been made to publish reliable data and information, but the author and publisher cannot assume responsibility for the validity of all materials or the consequences of their use. The authors and publishers have attempted to trace the copyright holders of all material reproduced in this publication and apologize to copyright holders if permission to publish in this form has not been obtained. If any copyright material has not been acknowledged please write and let us know so we may rectify in any future reprint.

Except as permitted under U.S. Copyright Law, no part of this book may be reprinted, reproduced, transmitted, or utilized in any form by any electronic, mechanical, or other means, now known or hereafter invented, including photocopying, microfilming, and recording, or in any information storage or retrieval system, without written permission from the publishers.

For permission to photocopy or use material electronically from this work, please access www.copyright.com (www.copyright.com/) or contact the Copyright Clearance Center, Inc. (CCC), 222 Rosewood Drive, Danvers, MA 01923, 978–750–8400. CCC is a not-for-profit organization that provides licenses and registration for a variety of users. For organizations that have been granted a photocopy license by the CCC, a separate system of payment has been arranged.

Trademark Notice: Product or corporate names may be trademarks or registered trademarks, and are used only for identification and explanation without intent to infringe.

Library of Congress Control Number: 2019952221

Visit the Taylor & Francis Web site at
www.taylorandfrancis.com

and the CRC Press Web site at
www.crcpress.com

Contents

About the Editors

Dr. Erik A. Beever is a Research Ecologist at the Northern Rocky Mountain Science Center of the U.S. Geological Survey and is Affiliate Faculty in the Ecology Department at Montana State University. Dr. Beever has nearly 100 publications in diverse scientific journals and in diverse subdisciplines of biology. He is a member of the IUCN World Commission on Protected Areas (where he serves as the North American Representative for the Mountains Network), the IUCN SSC Lagomorph Specialist Group, and the IUCN SSC Climate Change Specialist Group. He is also a member of The Wildlife Society (in which he chaired the Biological Diversity Working Group), Society for Conservation Biology, American Society of Mammalogists, World Lagomorph Society, Mountain Research Initiative, Consortium for Integrated Climate Research in Western Mountains, and Sigma Xi. He has served as a peer reviewer for 57 scientific journals, and delivered >200 presentations to local to international audiences. He has performed field research on plants, soils, most vertebrate clades (especially mammals), and insects, and in a range of ecosystems of the western hemisphere. In addition to investigating numerous aspects of disturbance ecology, he also seeks to understand mechanisms of biotic responses to long-term weather patterns and variability, and monitoring in conservation reserves, all at community to landscape scales, as well as other topics of conservation ecology, wildlife biology, and landscape ecology. He is interested in questions at the nexus of basic and applied science, especially those that also inform management and conservation efforts for species, communities, and ecosystems. After receiving his undergraduate degree in Biological Sciences at U.C. Davis, Erik received his Ph.D. from the Program in Ecology, Evolution, and Conservation Biology at the University of Nevada, Reno.

Suzanne Prange received her B.S. and M.S. in Biology from the University of South Alabama and her Ph.D. in Wildlife Biology from the University of Missouri-Columbia. She also completed post-doctoral training at Colorado State University and Ohio State University. The majority of her research has been dedicated to threatened and endangered forest wildlife species, and she worked extensively with the previously state-endangered bobcat in Ohio. In addition to bobcats, she has worked with numerous mammalian species, from southern flying squirrels to coyotes. She has authored peer-reviewed papers and book chapters, worked extensively on manuscript reviews for numerous journals, and served as an Associate Editor for the *Journal of Mammalogy*. She has served on several executive boards and committees within The Wildlife Society and the American Society of Mammalogists. Currently, she is dedicated to wildlife conservation research in Ohio's Appalachian region, and is the Executive Director and Lead Scientist at the Appalachian Wildlife Research Institute.

Dr. Dominick A. DellaSala is President and Chief Scientist of the Geos Institute (www.geosinstitute.org) in Ashland, Oregon and former President of the Society for Conservation Biology, North America Section (www.conbio.org). He is an internationally renowned author of over 200 science papers on forest and fire ecology, conservation biology, endangered species management, and landscape ecology. Dominick has given plenary and keynote talks ranging from academic conferences to the United Nations Earth Summit. He has appeared in *National Geographic*, *Science Digest*, *Science Magazine*, *Scientific American*, *Time Magazine*, *Audubon Magazine*, *National Wildlife Magazine*, *High Country News*, *Terrain Magazine*, *NY Times*, *LA Times*, *USA Today*, *Jim Lehrer News Hour*, CNN, MSNBC, *Living on Earth* (NPR), several PBS documentaries and even Fox News! Dominick is currently on Oregon's Global Warming Commission Subcommittee on Forest Carbon and is Editor of numerous scientific journals and publications. His book *Temperate and Boreal Rainforests of the World: Ecology and Conservation* (2011, Island Press: Washington, D.C.) received an academic excellence award from *Choice* magazine, one of the nation's top book review journals. His recent co-authored book—*The Ecological Importance of Mixed-Severity Fires: Nature's Phoenix* (2015, Elsevier: Boston)—presents groundbreaking science on the ecological importance of wildfires. Dominick co-founded the Geos Institute in July 2006 and is motivated by his work to leave a living planet for his two daughters, grandkids, and all those that follow.

Contributors

Gene Albanese
Kansas Cooperative Fish and
    Wildlife Research Unit
Division of Biology, Kansas State
    University
Manhattan, Kansas

Nicole Cavender
The Morton Arboretum
Lisle, Illinois, USA
Formerly The Wilds
Cumberland, Ohio

Henry Campa III
Michigan State University,
    Department of Fisheries and
    Wildlife
East Lansing, Michigan

Janet E. Foley
Department of Medicine and
    Epidemiology
School of Veterinary Medicine
University of California
Davis, California

David Haukos
U.S. Geological Survey, Kansas
    Cooperative Fish and Wildlife
    Research Unit
Kansas State University
Manhattan, Kansas

Alexandra Locher
Grand Valley State University
Allendale, Michigan

Benjamin T. Plourde
Department of Medicine and
    Epidemiology
School of Veterinary Medicine
University of California
Davis, California

David A. Pyke
U.S. Geological Survey
Forest and Rangeland Ecosystem
    Science Center
Corvallis, Oregon

Suresh A. Sethi
U.S. Geological Survey, New York
    Cooperative Fish and Wildlife
    Research Unit
Department of Natural Resources,
    Cornell University
Ithaca, New York

Kari E. Veblen
Utah State University
Department of Wildland
    Resources and Ecology Center
Logan, Utah

section one

Introduction

chapter one

Introduction

Defining and Interpreting Ecological Disturbances

Erik A. Beever, Suresh A. Sethi, Suzanne Prange,
and Dominick A. DellaSala

The Role of Disturbance in Ecology, Management, and Conservation

Disturbance plays a fundamentally important role in how ecological systems function and are structured, at multiple spatial and temporal scales. For example, disturbances may range from a single tree falling in a given moment (although the gap's successional trajectory can last for years to decades), to global climate change incrementally increasing springtime minimum temperatures. Ecological disturbance has been defined in many different ways over time, but in its simplest form it entails an externally forced change to a system that departs from the conditions that previously dominated. Contrary to the negative connotations that disturbance has outside the bounds of ecology, disturbances are not inherently positive or negative for ecosystems. Later in this chapter section *How can Disturbances be Categorized in Meaningful Ways?*, we review how characterizations of ecological disturbance have evolved over time. As described later in this chapter and throughout this book, disturbances and post-disturbance trajectories have profound implications for management and conservation of ecosystems and their components.

Two terms that have been inexorably related to disturbance ecology include *resilience* and *resistance* of a system to disturbance. Although resilience has blossomed into extremely wide usage and definitions in recent years (particularly in regard to contemporary climate change), historically it was defined in terms of the behavior of systems that are far from equilibrium, in particular as the magnitude of disturbance that a system can absorb without changing its state (Holling 1973). This meaning (and variants thereof) is now more commonly termed *resistance*. An alternative definition of *resilience* is the time that it takes for a system to return to an equilibrium or steady state following the disturbance (Tilman and Downing 1994; Neubert and Caswell 1997; Allison and Martiny 2008). This latter definition can be applicable either when a single, global equilibrium is believed to exist, or when multiple steady states are known to exist. Both definitions of resilience are strongly tied to the concept of system stability; in this book, we predominately use the latter definition.

Dynamics of disturbance have not only been the basis for much ecological theory, but have also intersected with numerous other topics of basic ecology. Perhaps the best-known instance of the former involves the Intermediate Disturbance Hypothesis (Connell 1978), which posits that species richness is highest in contexts and locations where disturbances occur with intermediate frequency. Classic examples of this disturbance effect include coral reefs subject to periodic intermediate disturbances that prevent dominance of predatory starfish that can greatly simplify community structure, and periodic fires of mixed intensities that create landscape heterogeneity associated with high levels of plant and animal richness (DellaSala and Hanson 2015). Although studies in a variety of contexts have provided evidence in support of the Hypothesis, approximately 80 percent of studies within meta-reviews have not supported the predicted peak of species diversity at intermediate intensity or frequency of disturbance (e.g., Mackey and Currie 2001; Fox 2013). Nonetheless, the heterogeneity created across landscapes by disturbances facilitates occupancy by numerous species (Sousa 1984), and disturbances can increase beta and gamma diversity by creating new microhabitats and making new resources available (Connell 1978; Lindenmayer 2009).

Starting from the classic perspective of successional theory (e.g., Clements' 1916 perspectives on climax states and facilitation), disturbance ecology has intersected with numerous realms of ecological theory to characterize post-disturbance recovery of ecosystems. More-recent ideas on succession that relate to disturbance ecology include Egler's (1954) Initial Floristic Composition model for abandoned agricultural land, hypotheses on Facilitation, Tolerance, and Inhibition (Connell and Slatyer 1977), and the Habitat Accommodation Model of Fox (1982). Other realms of ecological theory that relate to post-disturbance trajectories, which we describe at greater length below, include state and transition theory (e.g., Bestelmeyer et al. 2017), biological legacy theory (Franklin et al. 2000), and use of functional traits to predict species responses to disturbance (Vesk et al. 2004; Diaz et al. 2007). Although disturbance ecology has intersected with numerous other bodies of ecological theory, these have received extensive treatment elsewhere (e.g., Osenberg et al. 1999; Vilà et al. 2011; Gerstner et al. 2014; Thom and Seidl 2016).

Given the above-mentioned definitions of disturbance, evolutionary and co-evolutionary histories dictate (perhaps non-intuitively, if one assumes the negative connotations of disturbance also apply in ecology) that *lack of* natural disturbance can be more jarring than existence of regular disturbance, in certain cases. For example, some tree species with serotinous cones require fire or some other type of disturbance to scarify the seeds and subsequently have them germinate. Consequently, total suppression of fires in forested systems with such tree species will see those species decrease in abundance over time, particularly as the legacies of standing-dead trees leave the system. Similarly, in savanna and prairie ecosystems that co-evolved for continuous millennia with large, hooved mammals, removal of herbivory and fire can allow for such ecosystems to be invaded, simplified (by the rise in dominance of one or a few species), or otherwise degraded (Veblen et al., Chapter 4, this volume). Several species of cottonwood (*Populus* spp.) trees require scour of river banks from high-intensity flows before they can successfully germinate and establish. In these and other instances, scale and context can profoundly affect outcomes of disturbance. For example, an intense, large-scale natural disturbance in a forest with centuries-old trees often sets back succession to pioneering stages. If the interval between successive disturbances of large magnitude and intensity is comparatively short (relative to historical fire-return intervals; e.g., years to decades), the latter stages of succession are retarded. In some cases, managers and other conservation practitioners are experimenting with numerous techniques to try to mimic some aspects of natural disturbance regimes (e.g., prescribed fires of low intensity, instantaneous releases of large water volumes from dams).

Disturbances can be conceived as relating to the early work on the ecological niche, which is a species' role, distribution, position, diet, and behavior in its environment. In particular, because numerous selective forces (e.g., abiotic conditions and interactions with predators, prey, and competitors) act simultaneously on any organism, no organism is ever "perfectly adapted" (Fisher 1930, 1958). Fisher (1930, 1958) posited that an organism's realized level of evolutionary adaptation could theoretically be quantified if one were able to simultaneously assess evolutionary pressures on the organism in an infinite number of dimensions. Any change in the environment, be it biotic or abiotic, would thus constitute an ecological disturbance for organisms and, consequently, "deterioration of the environment" (Fisher 1930, 1958).

How can Disturbances be Categorized in Meaningful Ways?

Although disturbances vary along innumerable different axes, there are several classification schemes that may assist in categorizing disturbances in ways that are pragmatically or heuristically useful for investigating disturbance. Such schemes include characterizing disturbance: (a) as biotic or abiotic; (b) by the spatial and temporal scale(s) at which they occur; (c) in terms of the key characteristics of disturbances (timing, intensity, frequency, duration, and spatial extent); (d) as natural or anthropogenic; or (e) in terms of which aspects of biodiversity that they most strongly affect: structure, composition, or function of ecosystems. Arguably, characterizing a disturbance with respect to two or more of these classification systems provides richer and more-comprehensive information than characterizing a disturbance with respect to only one categorization scheme.

Biotic vs. Abiotic Disturbances

Perhaps the simplest way to categorize disturbances is to identify whether they are biotic or abiotic in nature. Anomalous, extreme weather conditions (e.g., severe drought, snow drought, flooding, heat waves, cold snaps, gale-force winds), other natural disasters (hurricanes, tsunami, earthquake, lightning), and pollution can constitute an abiotic disturbance for both natural and human communities. In contrast, invasive species, hyper-abundant native species (e.g., white-tailed deer *Odocoileus virginianus*), tree-fall gaps, diseases, pests, and herbivory constitute examples of biologically oriented disturbances. Given the complexity of natural (i.e., not anthropogenically dominated) ecosystems relative to the predictability of many chemical and physical laws, the dynamics of biotic disturbances tend to be more complex and challenging to manage and address restoration efforts for, but not always (Bestelmeyer et al. 2017; Lynch et al. *in press*). Nonetheless, feedback loops can exist in both biotic and abiotic disturbances (e.g., Beever and Herrick 2006). As most readers have probably experienced, magnitude, spatial extent, and frequency of disturbance can vary dramatically within both biotic and abiotic disturbances. Despite the apparent simplicity of this classification, some disturbances inherently involve both biotic and abiotic aspects. Although toxic algal blooms originate from altered water chemistry due to polluted runoff, the proliferation of algae in nearshore environs are the factor that can ultimately cause a cascade of ecological responses. Similarly, although wildfire is primarily an abiotic perturbation, the intensity, spatial extent, and frequency (e.g., fire return interval) of fires strongly depend upon conditions of the forest or other-vegetation mosaic.

Categorization Based on the Spatial and Temporal Scale(s) at which Disturbances Occur

A second way by which to categorize perturbations is with respect to the spatial and temporal scales at which they occur and have influence on ecosystems. Disturbance is pivotal in maintaining species diversity in many communities. Although the effects of disturbance frequency and extent on species diversity have been well studied, knowledge of the mechanisms of how the spatial structure of disturbance influences species diversity is still lacking (Liao et al. 2016). Modelling research has suggested that spatial characteristics of disturbance (e.g., extent, structure and pattern, position within the geographic range), as well as temporal aspects (e.g., frequency and intensity) have strong effects on population dynamics, which are often dependent upon the species (Hiebeler and Morin 2007; Banitz et al. 2008). Furthermore, many disturbances do not occur uniformly across the landscape (e.g., fire, drought, logging, flooding). The spatial effects of disturbance are therefore complicated and vary in multiple aspects.

Categorization Based on Key Characteristics of Disturbances (Timing, Intensity, Frequency, Duration, and Spatial Extent)

Third, disturbances may be categorized in terms of their key characteristics of intensity, spatial extent, frequency, and duration. Fires, floods, pest outbreaks, and species invasions, among other types of disturbance, affect many ecosystems over a range of spatial and temporal scales. These disturbance events have five key characteristics of particular importance: intensity, timing, duration, extent, and disturbance interval (Sousa 1984; Roxburgh et al. 2004). Together, these factors characterize disturbance regimes that are important to the composition and functioning of the affected systems (Dale et al. 1998; Sabo and Post 2008). For instance, a wildfire of mixed-severity effects on vegetation at the landscape scale creates a patchwork mosaic referred to as *pyrodiversity* that can be associated with extraordinary levels of biodiversity (DellaSala and Hanson 2015). The post-disturbance environment provides a habitat for a broad suite of species, from disturbance-adapted colonizers in severely burned areas to disturbance avoiders in fire refugia. The re-occurrence of natural disturbances of this nature perpetuates the pyrodiverse environment.

Natural vs. Anthropogenic Disturbance

As a fourth option, disturbances can also be characterized by whether they are anthropogenic (e.g., mining, forest logging) or natural (e.g., pest outbreaks, drought, flooding). In some cases, a disturbance may be either (e.g., prescribed vs. lightning-caused fires) and, in others, managers have begun to realize the importance of certain natural disturbances to ecological systems. Some disturbances have been reduced in frequency, intensity, or extent relative to that of the natural system, with resultant adverse effects on biodiversity. In certain cases, managers have used anthropogenic disturbances to mimic the natural ones. For example, cottonwoods support riparian zones and are vital to the success of the entire ecosystem by providing shade, shelter, and food. However, their reproduction is dependent upon riverine conditions. Many rivers in semiarid and arid environments have been dammed for water-management purposes, altering the natural variability of the river's flow and changing the amount of sediment carried by the river through the riparian areas. However, management of flow rates can be altered to improve cottonwood reproduction and recruitment, thereby protecting ecosystem functionality (Scott et al. 1997; Burke et al. 2009). Conversely, beaver dams constitute a natural disturbance that dramatically alters local hydrologic and vegetative conditions, and beaver ponds often increase beta biodiversity. In areas where beavers have been extirpated, reduced in density, or there is a high probability for beaver–human conflicts, human-produced beaver dam analog structures have been constructed to, at least in part, mimic beaver activity (Weber et al. 2017). Species and ecosystems have evolved many specialized adaptations to resist or be resilient to disturbances; many communities even thrive in the post-disturbance environment. For instance, natural disturbances often leave critically important structures of the pre-disturbance environment behind in their wake (i.e., biological legacies; see DellaSala, Chapter 3, this volume).

Categorization Based on which Ecological Components are Altered

Finally, disturbances can also be categorized depending upon the components of biodiversity (Noss 1990) that they affect (i.e., structure, composition, and function). Disturbances are natural features of ecosystems that promote variability in the community and ultimately maintain diversity. They create spatial and temporal variability in community structure in ecosystems, and the removal of species may allow competitively inferior ones to establish themselves, thereby altering community composition. Such local shifts in composition result in a patchwork of communities that can increase diversity at the landscape scale. Ultimately, the patchwork mosaic is affected by the frequency and intensity of disturbances, as well as spatial scale, and community resistance, or the ability to withstand a disturbance without altering ecosystem function.

Biological Diversity

Although biological diversity has existed as a concept in various forms since perhaps the time of Aristotle, it has only in the last half-century been recognized as a fundamental attribute of ecological communities in and of itself, termed "biological diversity" or "biodiversity" for short. Definitions of biological diversity have ranged from simply the number of species in a location (Schwartz et al. 1976; now termed species richness) to "… all of the diversity and variability in nature" (Spellerberg and Hardes 1992).

Franklin et al. (1981), Noss (1990), Niemi and McDonald (2004), and others have conceptualized biological diversity as a hierarchically nested, tripartite framework focused on structure, composition, and function. Such a characterization reflects the multiple spatial and temporal scales at which biological diversity is organized and interacts dynamically. The breadth of ecological components and processes included in the concept of biodiversity has become increasingly more comprehensive, including: species and their activities, genes, assemblages, biotic processes, ecosystems, other ecological components, and the interactions among them (DeLong 1996). More recently, biodiversity components and processes now also include behavior, phenology, life-history traits, gene expression, epigenetics, and ecological structure (at multiple scales), among other features (Sih 2013; Cutting et al. 2016; Beever et al. 2017; Hargreaves and Eckert 2019). Practically speaking, the complexity and comprehensiveness of the definition of biodiversity will depend on the context in which the concept is investigated, reflecting the relevant research questions and objectives. For example, the concepts of alpha, beta, and gamma diversity have been implemented as one means of characterizing biodiversity relative to a specific investigation, whereby alpha diversity describes local conditions, beta diversity compares diversity across local sites, and gamma diversity encompasses the sea- or landscape domain as a whole (Whitaker 1960). As another example, the concept of "evolutionarily significant units" (ESUs) is important for defining relevant genetic biodiversity as a conservation target; however, the definition of an ESU can initiate a chain reaction of policies and requisite management responsibilities (e.g., Pacific salmon *Oncorhynchus* spp. on the U.S. west coast; Waples 1995).

The recognition of biodiversity as a fundamental ecosystem attribute has advanced our understanding of the dynamics of ecological communities; however, the increasing comprehensiveness and nuance of how biological diversity is characterized has created practical challenges for management entities responsible for conserving biological diversity. In particular, such entities struggle in deciding which components (and how many components, collectively) to monitor, to index the broader, more-inclusive concept of biodiversity. Consequently, agencies as well as conservation-oriented non-governmental organizations (NGOs) have adopted a panoply of ways to index biological diversity with more-tractable measures that can be standardized, and thus compared across space and time. For example, numerous surrogate-species approaches have been proposed, including umbrella, flagship, keystone, focal, indicator, threatened, and endemic species, as well as functional guilds, common (as opposed to rare) species, and whole-community measures (Beever 2006; Caro 2010). Each of these approaches makes simplifying assumptions about how biological diversity is reflected in a given diversity index. In one such approach, management efforts targeted at "umbrella species" as a proxy for biodiversity presume that by conserving and tracking taxa that have spatially extensive home ranges and require high-quality, contiguous habitat, other species with narrower habitat niches and ranges will also be incidentally conserved under that "umbrella" (Roberge and Angelstam 2004, Caro 2010). Whole-community measures include everything from diversity indices (e.g., species richness, Simpson index, Shannon-Wiener index, Berger-Parker index, Magurran 2004), the index of biotic integrity (IBI; Karr 1981), the percentage of species that are native, amount of endemism, multivariate ordinations tracked through time (McCune and Grace 2002), to a litany of others. Other researchers have suggested using functional guilds, key ecological functions, and key environmental correlates as indices of biological diversity (Marcot et al. 1999). Ecoregions and "hotspots" have also been proposed as larger-scale measures of cumulative biological diversity. Additionally, some programs monitor habitat or land-cover types through time as a coarse index of which elements of biodiversity occur where.

In contrast with the above indices, however, striving to move from simply monitoring one or two charismatic ecosystem components (e.g., a particular bird, mammal, or tree species) to monitoring biological diversity more broadly acknowledges the multi-scale complexity and nature of ecosystems and espouses a more proactive approach to species conservation. Such an approach is likely to be a more effective and cost-efficient conservation strategy at longer time scales (Scott et al. 1995). In addition, monitoring biological diversity may better indicate ecosystem health than single-component approaches, and thus be more relevant for tracking the flow of ecosystem services that sustain human well-being. Furthermore, there is no single indicator species that will allow robust and specific inference to be drawn about the status of unmonitored species from the status of monitored species (Landres 1992). The need for plurality in monitoring ecosystems was also illustrated by a spatial reserve-selection approach for conserving species diversity across the Middle Atlantic region of the USA (Lawler et al. 2003). In that research, complete conservation of any single taxonomic indicator group (either freshwater fish, amphibians, reptiles, mammals, birds, or freshwater mussels) provided protection for only 17–58 percent of all other at-risk species (Lawler et al. 2003).

In applying ecological theory regarding patterns of biological diversity to ecosystem management, Bestelmeyer et al. (2003) suggest that conservation practice can be best informed by considering four questions: (1) which organisms or taxa are within the scope of the management problem?; (2) how do their scale-related domains relate to the relevant spatial scale of the management problem?; (3) which processes are likely to be important determinants of species distribution at management-relevant scales?; and (4) how will the proposed management actions interact with these processes? In efforts to understand the role of biodiversity (and its loss) to support conservation management and advance knowledge about the dynamics of ecosystems, ecological theory has been dedicated to describing patterns of diversity across space and time, particularly at meso- to macro-scales. Because this theory intersects at times with disturbance ecology, we provide entry points to the literature for key theories and conceptual frameworks related to biodiversity here, and refer the reader to these sources for further details. At the most fundamental level, five bodies of theory have been proposed to explain patterns of biological (particularly, species) diversity and distribution (Bestelmeyer et al. 2003a). These include macroecological theory (e.g., Rosenzweig 1995), gradient theory (e.g., ter Braak 1994), habitat selection and competition theory (e.g., Paine 1966; Wiens 1989b), spatial patch theory (e.g., Loureau and Mouquet 1999), and dispersal and vicariance-biogeography theory (e.g., Mayr 1963; Cracraft 1985). More specifically, patterns of biological diversity have been explained by the mid-domain effect (Colwell and Lees 2000), Island Biogeography Theory (MacArthur and Wilson 1963), stepping-stone dynamics, metapopulation dynamics, and numerous other ecological rules and hypotheses. Biological diversity has also been shown to correspond to latitude, habitat diversity and complexity, productivity, elevation, size of the continent or island on which the biota occur (and the island's distance from the mainland), niche requirements of the species involved, stability, and historical constraints such as phylogeny (e.g., Darwin 1859; Pianka 1994; Brown 1995; Rosenzweig 1995; Fine 2015; Ikeda et al. 2017). More recently, research into biodiversity as a fundamental process has expanded in scope, now looking towards the interaction of social and ecological systems (e.g., Leslie and McCabe 2013; Sethi et al. 2014; Folke 2016 and references therein).

Scale and Context Matter

How scale and context are addressed in disturbance ecology is fundamental to understanding biological organization, from genes to biomes and local to planet-wide processes. They are essential for setting the goals and objectives of research, management, and conservation approaches. Scales of disturbance may range from the falling of a single tree opening up a gap in a forest canopy, to continental-scale processes of plate tectonics affected by gravitational pull. Some authors (e.g., Barnosky et al. 2012) have argued that tipping points and state shifts due to various disturbances can be considered even at the planetary scale. A number of others have argued that it is valuable to sample at scales one higher and one lower than the primary scale of interest, to understand both cross-scale dynamics (Peters et al. 2004, 2007) and to contextualize the primary scale.

Spatial Scale

Numerous aspects of biodiversity, such as species richness, may be influenced by disturbance factors at scales as small as bare-soil patches that span one to several meters in diameter (Desoyza et al. 2000) to the structure of entire landscapes (Wickham et al. 1997; Mori 2011). Both wildfire and herbivory can produce gradients of disturbance effects at several different spatial scales, and the magnitude and direction of their influences on ecosystems vary widely across different extents. To adequately address the effects of herbivory, investigation at numerous spatial and temporal scales is required (Bisigato et al. 2005; Beever et al. 2006; Bisigato et al. 2008; Veblen et al., Chapter 4, this volume). Not only are movements of herbivores influenced by the heterogeneous nature of vegetation, soils, and weather dynamics (Milchunas and Noy-Meir 2002; Martin et al. 2018), but herbivore activity can modify the pattern of soils and vegetation at several spatial resolutions (Beever et al. 2018; Koerner et al. 2018).

Numerous analytical tools have arisen that allow for the explicit investigation of scale relative to any given ecological disturbance. Three of the most-broadly relevant such tools include variograms, semivariograms, and Moran's *I*. Stated simply, variograms describe the spatial continuity of the data. The experimental variogram is a discrete function calculated with a measure of variability over all pairs of observations with specified separation distance ("lag") and direction. Semivariograms graphically illustrate the degree of dissimilarity between observations (i.e., semivariance) as a function of distance. Moran's *I* is a correlation coefficient, ranging from −1 (perfect interspersion) to 0 (perfect randomness) to 1 (perfect clustering of similar values), that quantifies the overall spatial autocorrelation in a given data set (Moran 1950). Particularly for disturbances that may influence animal movements, resource selection functions (Manly et al. 2002) can use different radii around observed animal locations, to investigate which spatial scales are most relevant or predictive of dynamics. Hierarchically nested models (Beever et al. 2006; Albanese and Haukos, Chapter 7, this volume) are approaches that allow for different processes, factors, and nature of relationships to occur at distinct scales.

In addition to the above tools, numerous spatial-analysis modeling tools have arisen comparatively recently that allow investigation of the relationships of ecological disturbance to aspects of biological diversity. These are described in detail elsewhere (Fortin and Dale 2009), but include point-pattern analyses, mark correlation functions, network analyses of areal units, contiguous-units analyses, and circumcircle methods. A corresponding suite of particular statistical approaches bolster these methods, including joint count statistics, global (and local) spatial statistics, interpolation in spatial models (wherein patch and boundary dynamics are involved) and spatial-temporal analyses (Fortin and Dale 2009). In addition to these techniques, Hierarchical Bayesian Models (including state-space models) may also prove valuable for investigating ecological disturbance across space and time. Such models may be well suited because they can: accommodate sparse data across time or space, account for uncertainty in the observations, allow combination of multiple data sources and model structures, and provide estimates of the mean and variability across scales (Kéry and Schaub 2012; also see the chapter section *Advances in our Understanding of Disturbance Ecology*).

Temporal Scale

Temporal patterns (in either resolution or extent) have also been shown to dramatically affect the magnitude and direction of responses to disturbance. In habitat-selection investigations for example, variation in seasonal and annual phenomena can generate patterns that are misunderstood or overlooked when examined from an inappropriate temporal extent or resolution (Boyce 2006). Interpretation of population dynamics can also change dramatically, depending on the years and duration in which a species is investigated, particularly for species that exhibit population cycles (e.g., Brady and Slade 2004). Disturbances can affect the same plots or the same organisms quite differently in different seasons or years (Teague et al. 2004, Beever et al. 2017), often as a result of diverse weather conditions. Different time lags to the same disturbance in responses of diverse ecosystem components can also produce temporally inconsistent understandings of a given disturbance. Typically, heterogeneity across time in response to either disturbance or restoration efforts can only be solidly established after a long time series of data has been collected.

Context

For numerous aspects of ecosystem-disturbance dynamics, *context* also matters (e.g., Mori 2011; Hällfors et al. 2016; Ikeda et al. 2017). For example, context matters for the influences of the broader landscape and how it can modulate outcomes of a given disturbance in a smaller area. This may happen via dispersal, recolonization (e.g., from regional species pools), edge effects, and flows of water, sediment, and nutrients into the smaller area. Conversely, consequences of a disturbance in a discrete localized patch can ripple outward as a function of the patch's isolation from the broader habitat mosaic, the ecological integrity and resilience of the mosaic, and other factors. With respect to herbivory disturbance in particular, impacts of herbivores on ecosystem components are heavily context-dependent. Consequences of herbivory vary for example due to topographical position (Fowler 2002), the species of herbivore (Warner and Cushman 2002), timing and duration of the herbivory, presence of other interacting disturbances (Hobbs 2001; Harrison et al. 2003; Beever et al. 2008), levels of water, soil nutrients, and light (Hawkes and Sullivan 2001), and the ecosystem component being disturbed by the herbivory

(Milchunas et al. 1998; Jones 2000). Of course, ecosystems exist within a matrix of ubiquitous human activity in the Anthropocene (i.e., humanity's increasing ecological footprint), and thus disturbance impacts are modulated by the presence of anthropogenically related factors (e.g., non-native species, habitat fragmentation or degradation, occurrence of other disturbances), as well as the biophysical template or "stage" on which various species "actors" will come and go, as environmental and climatic conditions change (Brost and Beier 2012).

Further illustrating the importance of context, with respect both to niche theory and to how contemporary climate change is affecting populations of various species, the position of a population location within a species' geographic range can influence the impact of disturbance regimes. From a theoretical standpoint, many authors have argued using niche theory that abundance of species should be highest at the range center and decrease as one moves towards the range margins (i.e., the Abundant-Center Hypothesis; Andrewartha and Birch 1954; Hengeveld and Haeck 1982). Tests of this paradigm have provided mixed results, however, partly because biotic and abiotic conditions do not shift monotonically away from any particular value that is optimal for a given species (Sagarin and Gaines 2002; Sagarin et al. 2006; Samis and Eckert 2007). Biotically, such unpredictable nonlinearities result from, among other factors, vicariant events that allow dispersers to end up in some areas but not others, purely by chance. Abiotically, such complexities arise from irregular and complex physiography, cold-air pooling, existence of microrefugia, and decoupling of microclimate from meso- and macro-climates, among other phenomena (Dobrowski 2011; Varner and Dearing 2014).

Given the multi-scale nature of disturbances (Perry and Amaranthus 1997; Wickham et al. 1997) generally and of many disturbances in particular, as well as the multi-scale nature of many ecological components and processes (Wiens 1989a; Bestelmeyer et al. 2003a), disturbance effects on ecological communities can be quite context-dependent. For example, disturbance effects on species diversity are inherently scale-dependent, because the relevant spatial scale depends critically on whether one is interested in understanding patterns in alpha, beta, or gamma diversity (Whittaker 1960). Wildfire and herbivory (discussed in Chapters 3 and 4, this volume) have attracted research attention across numerous scales. Disturbances such as these are heterogeneous across space and time, and such heterogeneity is compounded because landscapes possess varying levels of vulnerability to disturbance (Perry and Amaranthus 1997; Bestelmeyer et al. 2003a). Additionally, whereas some landscapes typically absorb and dampen the spread of disturbances, other landscape patterns can magnify their spread (Peters et al. 2004, 2007).

Advances in our Understanding of Disturbance Ecology

As collective empirical evidence and analytical advances accumulate, our view of disturbance ecology has evolved considerably (e.g., Pulsford et al. 2016). Fundamentally, disturbance ecology seeks to understand the impacts of, and ecosystem response to, shocks (i.e., perturbations). By the turn of the millennium, a useful framework had arisen for categorizing the nature of disturbances as either "press"—chronic perturbations experienced by an ecosystem over a long time period—or "pulse" phenomena—acute shocks experienced over discrete time periods (e.g., Stanley et al. 2010; Pulsford et al. 2016). Since this time, press and pulse disturbances are now viewed as bookends for a continuum of perturbations for which ecosystem responses are driven by the frequency, magnitude, and direction of shocks (e.g., Benedetti-Cecchi et al. 2006; Donohue et al. 2016). One alternative characterization in particular along this continuum is "ramp" disturbances; these are characterized by the strength of the disturbance increasing steadily over time, often accompanied by simultaneous increases in spatial extent (Lake 2000). Ramp disturbances such as droughts or increasing sedimentation of a stream after the forest cover in its catchment has been cleared may continue increasing without an endpoint, or may reach an asymptote after a long period of time.

Conceptual understanding about the mechanisms linking disturbances to ecosystem responses continue to advance and the complex and nonlinear dynamics exhibited by systems are now believed to be widespread (e.g., Scheffer et al. 2001; Foley et al. 2015). As a result, considerable research has been devoted to describing dynamics of systems as a means to predict the future trajectory of ecosystems under disturbance regimes. These models are improving our understanding about extreme changes, because disturbed ecosystems can suddenly pass tipping points and jump to new system dynamics, although our forecasting abilities remain weak (e.g., Scheffer et al. 2009; Sugihara et al. 2012). Mechanistic understanding about the attributes of systems that lead to resilience continue to advance, such that the field has now moved beyond a simple view of a diversity–stability relationship towards one that is built from functional traits (e.g., Heilpern et al. 2018; Dakos et al. 2019). Furthermore, the scope of disturbance ecology has expanded to explicitly consider both social and ecological systems as linked in determining ecosystem dynamics (e.g., Collins et al. 2011).

An important advance in the theory underlying disturbance ecology is recognition of context dependencies, whereby similar disturbance mechanisms can drive different ecosystem responses depending on legacies of biological communities, lagged response dynamics, geographic scale, and feedbacks among system stressors (e.g., Shears et al. 2008; Chamberlain et al. 2014; González-Moreno et al. 2014). For example, DellaSala (Chapter 3, this volume) finds that differing legacies of fire and logging drive forest context, which in turn mediates tree mortality and patch dynamics in Pacific Northwest forests. Context dependency presents complications for the generality of disturbance-ecology insights, such that results may not transfer well across case studies, owing to context. Nevertheless, recognition of context dependency has refocused ecological research priorities towards deeper understanding of the interactions between fundamental ecological processes and system context (e.g., Agrawal et al. 2007).

Complementing theoretical advances in our understanding of disturbance ecology, novel analytical and technological tools have improved our ability to assess ecosystem disturbance dynamics across scales of space, time, and ecological complexity (Mori 2011). With increases in computing power, two important analytical advancements have expanded the statistical tools available to monitor spatiotemporal patterns in population- or community-level processes that are useful for assessing system disturbance and response dynamics: hierarchical ecological modeling and critical transition (e.g., tipping points, regime shifts, ecological thresholds) detection analyses. Hierarchical statistical models separate out the true but unobservable underlying ecological phenomenon of interest (e.g., abundance over time or distribution over space) from observation errors, which reflect our imperfect ability to measure events and collect data, and from natural process error. At the level of individual species, a number of hierarchical statistical models have recently been developed that explicitly separate detection processes from true underlying abundance and distribution data, improving our ability to assess interactions between organisms and their environments (Royle and Dorazio 2008). These models (e.g., distance sampling, occupancy modeling, and N-mixture modeling; see Kéry and Royle 2015) allow for unbiased assessment of spatial ecology across different types of habitats, where naïve observations would lead to conflation of the difficulty in observing individuals and the species' true distribution (MacKenzie et al. 2002). Furthermore, "dynamic" occupancy models extend these approaches to consider colonization and extinction processes and have proven useful for understanding metapopulation dynamics of populations facing environmental disturbances (e.g., Govindan et al. 2011; Calvert et al. 2018). Hierarchical ecological models that control for detection processes are particularly well suited for eDNA sampling, which is providing new tools with unprecedented sensitivity to detect the presence or assess the abundance of species (Carreon-Martinez et al. 2011; Sethi et al. 2019) and ecological communities, particularly in aquatic environments (e.g., Evans et al. 2016).

Beyond single-species hierarchical models that control for detection, state-space modeling has advanced our ability to address temporally or spatially correlated ecosystem processes. State-space models are a general class of autoregressive hierarchical statistical estimators that include linked component models for underlying ecological state dynamics, with associated process and observation error models (e.g., de Valpine 2002; Pedersen et al. 2011). Because the models are dynamic, insofar that the value of a process of interest at one point in time or space is a function of the value at a previous point in time or space, they can accommodate spatial and temporal correlations in both the state and observation processes. This may be of interest in assessing the spatiotemporal scales of ecosystem disturbances and responses. In fact, dynamic occupancy modeling can be viewed as a state-space model (Royle and Kéry 2007). Other forms of state-space models have been used to assess the response of tree species to climatic disturbances (e.g., Hurteau et al. 2007) or to assess the spatial ecology of ungulates in relation to heterogeneous landscapes (e.g., Forester et al. 2007), among other applications.

Vector-based (also referred to as "multivariate") state-space approaches extend these approaches to simultaneously model the space-time trajectories of multiple entities and thus provide opportunity to estimate the nature and strength of interactions among ecological-community components with each other or in response to environmental drivers. These approaches account for spatiotemporal correlation processes and the challenges common with ecological data, such as observation errors and partial data coverage (e.g., Holmes et al. 2012). Multivariate state-space models have also been used to assess best practices for analysis of baseline ecological indicators for marine infrastructure development (offshore energy) disturbances on benthic ecosystems (Linder et al. 2017), to explore the relationship between rainfall drivers and small mammal communities (Greenville et al. 2016), and tease apart biotic and abiotic drivers of phytoplankton communities (Baraquand et al. 2018). Additional applications are quickly increasing in number.

The ability to detect and forecast critical transitions as systems shift to novel dynamic behavior is a key analytical approach needed to assess and manage the response of ecosystems as climatic and other disturbances accumulate (Scheffer et al. 2001). Given that nonlinear dynamics and high variability are commonplace in ecosystems, such forecasting is difficult in practice. A number of statistical approaches have emerged to detect when critical transitions have occurred in time series of data on ecosystem components (e.g., Andersen et al. 2009). These typically involve sequential tests over moving windows of data, or comparisons of variances and trend pre- and post-candidate "change points" that indicate when the behavior of a system has changed (e.g., Reeves et al. 2007; Killick and Eckley 2014). Although the detection of critical transitions may be possible once they have occurred, accurate forecasts of critical transitions for complex systems under stress have remained elusive. Recent work has found commonalities in "early-warning" signs prior to critical transitions across complex systems as disparate as financial markets and ocean ecosystems (e.g., May et al. 2008; Dakos et al. 2012; Litzow et al. 2013). Empirical evidence has demonstrated that strongly changing nonlinear components of systems tend to increase in variance prior to critical transitions, and autocorrelation among interrelated features also tends to increase (e.g., Scheffer et al. 2009; Foley et al. 2015). Similarly, systems can slow down in the ability to recover from disturbance or begin to "flicker" between alternative states. Research into detecting and predicting the behavior of dynamic systems remains an active area of work. These tools will be critical for assessing the degree to which ecosystems have responded to disturbance regimes and inform the options available for managers to attempt to resist dramatic change or to actively direct its course in disturbed ecosystems (e.g., Lynch et al. *in press*; Thompson et al. *in press*).

Analytical tools to assess ecosystem disturbance dynamics are rapidly expanding; however, advancements in disturbance ecology ultimately rely on data. Complementing analytical advances to inform disturbance ecology, remote-sensing tools are now providing spatiotemporally extensive datasets to observe landscape features at unprecedented spatial scales. Recognizing that these data streams require considerable technical capacity to process and analyze, satellite imagery has been used to assess changes in terrestrial plant communities at broad spatial scales in both the near and longer terms in response to drought (e.g., Verbesselt et al. 2012) or climate-driven pest outbreaks in forest stands (e.g., Coops et al. 2010). Artificial intelligence tools (e.g., Olden et al. 2008) are further extending the applications of satellite imagery to recognize specific landscape features as a means to track ecosystem responses to disturbances such as deforestation (Jakimow et al. 2018).

Although the tools and data streams available to investigate the complexities of land- and seascape disturbance dynamics continue to improve, managers' needs for decision-support tools have expanded as broad-scale disturbances and increasing anthropogenic demands on ecosystem services have mounted. The "state-and-transition" framework has provided a pragmatic approach for communicating and modeling spatially and temporally explicit landscape outcomes under different disturbance regimes (e.g., Bestelmeyer et al. 2011; Daniel et al. 2016). The framework and underlying theory cast ecosystems as following a suite of possible states whereby locations on the landscape transition in a stochastic fashion from one form ("state") to the next over time as a function of disturbances or natural ecosystem trajectories (e.g., stand growth). A suite of anthropogenic and natural disturbances can be added into these models, providing analysts and managers with a tool to simulate landscape-level responses to different management strategies that address ecosystem disturbances (e.g., deforestation, urbanization, drought, fire, etc.). A key feature of state-and-transition models is that ecosystems need not follow a regular, pre-determined order through states, e.g., as would occur in traditional Clementsian (Clements 1916) succession, and thus can represent complex responses and non-equilibrium dynamics expected in disturbed landscapes. The state-and-transition framework has existed for some time (e.g., see Westoby et al. 1989; Briske et al. 2005 and references therein); however, its implementation has recently become more widespread, particularly in rangeland management (e.g., Bestelmeyer et al. 2003b) and, more recently, aquatic-ecosystem management (e.g., Zweig and Kitchens 2009; Bond et al. 2019).

Extreme warming events (Rahmstorf and Coumou 2011), precipitation extremes (Coumou and Rahmstorf 2012), increasing impacts of invasive species (Early et al. 2015), and continued terrestrial habitat conversion (Steffen et al. 2015), among other anthropogenic stressors, now reach ecosystems throughout the globe (Whitmee et al. 2015). Although theoretical and analytical advances have improved our ability to describe and model dynamics of ecosystem disturbance, this increasing frequency and extent of anthropogenic disturbances has led to ecosystem conditions that lack historical analogs (e.g., Nolan et al. 2018). In some cases, it has also led to entirely new combinations of biotic actors and abiotic conditions (Williams and Jackson 2007; Hobbs et al. 2009). As these novel ecosystem configurations continue to develop under anthropogenic disturbance regimes, we will have little empirical evidence or existing experience with understanding these new systems (also see Cavender and Prange, Chapter 6, this volume). Thus, as the field of disturbance ecology continues to evolve, a focus on understanding the underlying mechanisms with which disturbances drive ecosystem structure and function will be critical for forecasting and managing impacts of disturbances in these transformed ecosystems (e.g., Lynch et al. *in press*; Thompson et al. *in press*).

Four Themes Woven Throughout this Book

In this book, we have chosen to emphasize, across all of the chapters, four themes that are integral to disturbance ecology:

Theme 1: Ties to mechanisms related to fundamental principles or theory of ecology, including (when appropriate) the role that thresholds play in such systems;

Theme 2: Context-dependence and unpredictability (e.g., multiple possible pathways; nonlinearity);

Theme 3: Importance of antecedent conditions and past disturbance legacies; and

Theme 4: Relationship of natural and anthropogenic disturbances.

In addition to these four core themes that underpin the remainder of this book, the subsequent chapters are organized into three subject areas relevant to contemporary disturbance ecology, which we list in the rest of this paragraph. Given that disturbance ecology has had a rich and long history in forested ecosystems, Section II (including chapters by Campa and Locher, and DellaSala) focuses exclusively on forested ecosystems, and addresses issues of scale-dependence, natural vs. anthropogenic disturbance, and length of recovery or succession trajectories. Section III (chapters by Veblen et al., Foley and Plourde) addresses herbivory and emerging disturbances amidst global change, focusing especially on non-native species, disease, and synergies with other disturbances. The final Section (Cavender and Prange, Albanese and Haukos) focuses on land-use disturbance, exploring in particular the topics of landscape pattern, resilience, and recovery.

Chapter 2: Wildlife Responses to Abiotic Conditions, Herbivory, and Management of Aspen Communities

The loss of natural disturbance, including the loss of large carnivore predation on herbivores, can lead to "cascading trophic effects" as exemplified by predator–prey dynamics in aspen (*Populus* spp.) communities. For instance, Campa and Locher (Chapter 2, this volume) discuss the complexities of maintaining aspen (legacy) in Michigan where heavy ungulate herbivory has resulted in aspen-regeneration failures. Adjusting ungulate hunting pressure and changes in forest management are proposed to restore aspen. A similar situation of intense browsing pressure from ungulates (Rocky Mountain elk, *Cervus canadensis*) also existed in the Greater Yellowstone region prior to gray wolf (*Canis lupus*) reintroduction in the early 1990s. Shortly after reintroduction of gray wolves, aspen returned to the Park as elk numbers dropped and elk distributions moved increasingly away from predator pressure inside the Park (e.g., before reintroduction, many elk spent extended time periods in riparian corridors, in several drainages). Thus, large carnivore restoration in the park affected the entire aspen community along with shifting herbivore distributions to a more-natural dynamic.

Chapter 3: Fire-mediated Biological Legacies in Dry Forested Ecosystems of the Pacific Northwest, USA

DellaSala (Chapter 3, this volume) demonstrates how antecedent conditions are connected to post-disturbance dynamics through the presence of biological legacies at multiple spatiotemporal scales. Furthermore, these connections, in part, are a common way in which natural disturbances differ from anthropogenic ones. For instance, in the Pacific Northwest, natural disturbances range from disease affecting individual tree branches to tree mortality within a given forest stand to landscape-scale fires of mixed-severity effects on vegetation. The quality of biological legacies at the stand and landscape levels (i.e., context matters) depends on the pre-disturbance environment (antecedent conditions), which then sets in motion the sequence of forest-community stages and degree of biodiversity present in them through time. In this sense, forest succession is not a linear process (from young to old), but rather a series of linked and circular events that cycle through a forest over time.

Managers can enhance stand- and landscape-scale structures by retention of biological legacies (wherein quantity, quality, scale, and context matter), and by managing for pyrodiversity (diversity of fire patches) produced by large fires at broader scales. Wildfires could be managed safely for forest ecosystem services via minimally invasive fire suppression tactics in many low-use landscapes such as roadless areas or wilderness areas. Additionally, postponing or avoiding post-fire logging and tree planting immediately following high-intensity burns that kill most of the trees on site can allow for the expression of complex early forests that are known to be as biodiverse as old growth yet are presently rare. In addition to fire management, harvest practices can be managed to promote forest community diversity. For example, lengthening the time between successive timber harvest rotations in working forests typically allows for development of older structures that could be retained between harvest cycles.

Chapter 4: Context-dependent Effects of Livestock Grazing in Deserts of Western North America

Veblen et al. (Chapter 4, this volume) provide an overview of grazing by large mammalian herbivores as a disturbance process that involves selective removal of herbaceous plants, redistribution of nutrients via defecation and urination, creation of patch mosaics that differ in grazing intensity (i.e., increased heterogeneity), and hoof action on plants and especially soil surfaces. Veblen et al. argue that the ecological consequences of grazing are strongly context-dependent, varying with respect to numerous factors. These factors include coevolutionary history of plant–herbivore dynamics in the area (such that Great Plains and savannah systems respond more positively to grazing pressure than do deserts), timing and duration of the grazing, species and density of other large mammals in the vicinity, biophysical context (e.g., cooler, wetter ecosystems typically respond more favorably to grazing), landscape position (e.g., potential for water run-off inputs from higher elevations, northern vs. southern aspect), geology, soil parent material, and topography, among others.

Legacies of historic high-intensity grazing by domestic sheep and cattle in western North America during 1890–1920 continue to leave signatures such as degraded soil conditions, depauperate native-plant communities, and lower resilience to present-day grazing disturbance. In managed grazing systems, fenced livestock are often kept at stocking rates higher than what existed across the year for native herbivores, and native herbivores often move more quickly (i.e., within hours to weeks) through a given portion of the landscape than do domestic animals. The effects of grazing are complicated by contemporary climate change, in that hotter and effectively drier locations are typically less resilient to grazing. Grazing has been linked to fundamental ecological theory by numerous tests of the Intermediate Disturbance Hypothesis, a topic discussed earlier in this chapter. Results have been mixed. For instance, experimental tests on the northern slope of Mt. Kilimanjaro (where plants co-evolved with numerous ungulate species) and in high-productivity meadow–steppe communities in northeastern China found richness or standing biomass was highest at intermediate stocking rates. However, a test in the western Great Basin Desert found that as intensity of grazing disturbance by free-roaming horses increased, richness of small mammals, plants, and ant mounds monotonically decreased. Veblen et al. discuss how shrub encroachment and invasive species alter disturbance dynamics dramatically (e.g., via more-frequent fires), and review how the consequences of *removing* grazing are also highly context-dependent. The chapter ends with an in-depth case study investigating the effect of removing livestock and feral burros from Mojave National Preserve on several plant and soil response variables.

Chapter 5: Response of Tick-borne Disease to Fire and Timber Harvesting: Mechanisms and Case Studies across Scales

After mosquitoes, ticks are the most important zoonotic vector of human and animal diseases worldwide (Yu et al. 2015). Many tick-borne diseases are maintained by vertebrate hosts, and tick vectors can be severely affected by disturbance. Two disturbances that occur repeatedly, have important impacts on tick-borne disease, and are relatively under-studied in this context are fire and timber harvesting. Whereas fire may be natural or anthropogenic in nature, timber removal at the typical scales of harvest (e.g., stand to hillside) is decidedly the latter. Foley and Plourde (Chapter 5, this volume) present four hypotheses regarding disturbance and tick-borne disease. One, "the cruel forest," predicts that intact, heterogeneous ecosystems provide all the ecological needs of pathogens and vectors, and habitat disturbance could reduce disease risk. Second, "the kind forest" is also heterogeneous and supports a diversity of hosts, but these maintain a low disease prevalence and disturbance could increase risk. Third, "the simple forest" is composed of a more-homogeneous habitat with low biodiversity, and moderate disturbance would decrease disease risk. Finally, in "the hidden forest," the natural ecosystem is more isolated from people and domestic animals, and disturbance may remove buffers and increase disease risk. Upon examination of tick-borne disease responses to disturbance, Foley and Plourde find support for all four hypotheses, thus illustrating the multiple possible pathways or responses of disease dynamics to disturbance and the importance of antecedent conditions (e.g., homogeneous vs. heterogeneous habitat). Furthermore, the true effects of disturbances on disease risk depend on variation in space, time, disturbance intensity, and ecological contingency (i.e., are context-dependent). Unique components of different ecosystems lead to different and nonlinear responses to similar disturbances.

Foley and Plourde also address areas where additional study is warranted to promote an understanding of the ecological mechanisms and contexts that determine the change in tick-borne disease risk and disturbance. Many factors affecting tick-borne disease outcomes following disturbance have received sparse attention, including study across spatial scales, habitat types, disturbance intensities, and the spatial extent of disturbance, as well as at remote sites versus those adjacent to already-disturbed sites. Research is also lacking at fine scales for timber harvesting, all scales for high-intensity fires, and regarding detailed mechanisms that link disturbance through a causal pathway to ticks and disease risk. Realistically, however, ecological contingency is critically important in vector-borne disease response to disturbance, and accurate prediction will likely remain elusive. Nevertheless Foley and Plourde argue that increased knowledge will allow us to increase our understanding of changes in tick-borne disease risk across both space and time in disturbed landscapes.

Chapter 6: Microbes to Bobcats: Biological Refugia of Appalachian Reclaimed Coal Mines

The Appalachian region is one of the most biologically diverse areas in the temperate world and provides important ecosystem services, including storing carbon and maintaining watershed integrity and water quality (Olson and Dinerstein 2002). Nonetheless, Appalachia is severely affected by anthropogenic disturbances, from agriculture to timber harvest to surface mining. Over 600,000 hectares of Appalachian forests were mined between 1978 and 2009, and an additional 10,000 hectares are mined annually (Zipper et al. 2011). The process of surface mining alters the physical, chemical, and biological characteristics of ecosystems by removing existing vegetation, altering the soil profile and chemistry, and displacing or destroying wildlife, among other effects. Cavender and Prange (Chapter 6, this volume) describe a case wherein reclamation laws improved environmental quality, including increased erosion control, improved water quality, amelioration of extreme pH, and improved stability of topsoil through removal of vertical exposed walls of soil. However, additional ecological challenges were created—grading equipment caused significant compaction of the soil, resulting in low soil porosity, permeability, and moisture-holding capacity. By law, sites need not be restored to their original ecosystem, and many native forests have been replaced with exotic grasslands. Natural grasslands are one of North America's most-threatened habitat types, as the great majority of its grasslands have been converted to agricultural lands for crops and livestock. Thus, reclaiming mineland to prairie could represent refugia for many species declining because of grassland habitat loss, while potentially allowing a slow recovery to native forest. Reclamation, when done properly and coupled with active management, could reinitiate natural ecosystem processes; thus, reclamation is a valuable tool to mediate the large-scale anthropogenic disturbance of mining and conserve the countless ecosystem services of the region.

The Wilds is a 3705-hectare international wildlife ecotourism and conservation center located in the Appalachian ecoregion in southeastern Ohio. The Wilds has experienced substantial disturbance from surface mining for coal, although some remnant forest patches still remain; over 90 percent of The Wilds is in various stages of post-reclamation or restoration. The area lends itself well to scientific study of recovering ecosystems and wildlife and the alternate pathways of recovery that can result after human disturbance. As a case study, there are several factors that are likely responsible for the terrestrial ecosystems of The Wilds serving as successful habitat. First, despite much of the area being grassland, remnants of original hardwood forest remain, as well as forests reclaimed prior to the 1977 reclamation law implementation, which signify the importance of legacy conditions. These forest remnants likely serve as refugia that enhance re-colonization of surrounding disturbed areas, as well as improve overall landscape connectivity. Information observed over decades from biological communities, including microbes (bacteria), cellular slime molds (dictyostelids), fungi, gastropods, vegetation, birds, pollinators, small mammals, and predators have provided insight into biodiversity and its relation to multiple disturbances, habitat, recovery, succession, spatial distribution of habitats, management regimes, presence of refugia, and overall ecosystem functioning. Of particular interest is ecosystem functioning and wildlife recovery following mining and subsequent reclamation, and management practices that can promote stable, functioning ecosystems in these areas. Previously-mined lands are often dismissed because of their history of disturbance; however, there is great potential for reclaimed mines to be important areas for the conservation of many trophic levels, as well as for the structure and function of biological diversity.

Chapter 7: Toward a Theory of Connectivity among Depressional Wetlands of the Great Plains: Resiliency to Natural and Anthropogenic Disturbance within a Wetland Network

As illustrated in Chapter 7, disturbance across temporal and spatial scales is integral to the function of wetland systems. Albanese and Haukos (Chapter 7, this volume) use depressional, inland wetlands of the Great Plains as a model system to explore network dynamics and how they are affected by sediment accretion (and thus wetland infilling). These systems, also known as prairie potholes or (if drier) playas, are described using network analysis to quantify the contribution of individual wetlands to numerous system-wide aspects of biological diversity (e.g., ecosystem function, migratory birds, redundancy and resiliency, etc.), at multiple spatial and temporal scales. The authors argue that although regulatory and management frameworks are focused on stability, resiliency, and function of individual wetlands, many components of biological diversity (e.g., migratory birds) respond equally or more strongly to the mosaic of conditions across the entire interdependent network. Albanese and Haukos (Chapter 7, this volume) also note that whereas natural inundation of individual wetlands is stochastic and patchy within and across years, many anthropogenic activities that contribute to loss of topsoil are predictably leading to increasingly infilled depressions from accretion. They propose modeling dynamics with a spatially explicit, topological network hierarchical (i.e., nested) model that is grounded in the interactions between disturbance and dispersal processes and their constraints across scales. In their chapter, thresholds are expressed in terms of scale, such that certain dynamics are evident only beyond a particular scale. Network analyses can be used in such systems to rank and prioritize the contribution of individual playas by combining spatial distribution, disturbance frequency (i.e., inundation), and ecological condition. Albanese and Haukos conclude that conservation management of these complex, diverse interconnected systems is best informed by examining both fine- and broad-scale dynamics simultaneously; historically, the latter type have received much less attention.

Concluding Remarks

Although we explore numerous topics in this book, many research frontiers that will profoundly affect applied management and conservation remain. Each of these has attracted some attention in the literature to date, yet deeper understanding from a wider range of contexts is likely to better inform decision-making and thinking on the topic. For example:

- What might be some ways of grouping disturbances to help address strategies for conservation and management?
- Do natural disturbances form a framework that can be used in evaluating anthropogenic disturbances?
- How can disturbance scales be evaluated (singly, or simultaneously) and used in restoration and management strategies?
- Are responses at community or species levels different in frequently disturbed systems compared with more "stable" systems?
- How does phylogeny influence disturbance responses? For example, is response to land-use disturbance (e.g., fragmentation) dependent on dispersal capacity, and if so, at what scales?
- If natural disturbances are considered key processes to maintain biodiversity (an oft-held view), how should conservation and management responses to these disturbances be considered?
- How can the complex biodiversity associated with natural disturbances be mimicked in increasingly anthropogenically dominated land- and sea-scapes?

These and many other topics will provide interesting fodder for continuing investigation and catalyze wider and deeper understanding of how disturbances shape the biological diversity that characterizes planet Earth.

References

Agrawal, A. A., D. D. Ackerly, F. Adler, A. E. Arnold, C. Cáceras, D. F., Doak, E. Post, P. J. Hudson, J. Maron, K. A. Mooney, M. Power, D. Schemske, J. Stachowicz, S. Strauss, M. G. Turner, and E. Werner. 2007. Filling key gaps in population and community ecology. *Frontiers in Ecology and the Environment* 5: 145–152.

Allison, S. D., and J. B. H. Martiny. 2008. Resistance, resilience, and redundancy in microbial communities. *Proceedings of the National Academy of Sciences of the United States of America* 105: 11512–11519.

Andersen, T., J. Carstensen, E. Hernández-García, and C. M. Duarte. 2009. Ecological thresholds and regime shifts: approaches to identification. *Trends in Ecology and Evolution* 24: 49–57.

Andrewartha, H. G., and L. C. Birch. 1954. *The Distribution and Abundance of Animals*. Chicago: University of Chicago Press.

Banitz, T., A. Huth, V. Grimm, and K. Johst. 2008. Clumped versus scattered: how does the spatial correlation of disturbance events affect biodiversity? *Theoretical Ecology* 1: 231–240.

Baraquand, F., C. Picoche, D. Maurer, L. Carassou, and I. Auby. 2018. Coastal phytoplankton community dynamics and coexistence driven by intragroup density-dependence, light and hydrodynamics. *Oikos* 127: 1834–1852.

Barnosky, A. D., E. A. Hadly, J. Bascompte, E. L. Berlow, J. H. Brown, M. Fortelius, W. M. Getz, J. Harte, A. Hastings, P. A. Marquet, N. D. Martinez, A. Mooers, P. Roopnarine, G. Vermeij, J. W. Williams, R. Gillespie, J. Kitzes, C. Marshall, N. Matzke, D. P. Mindell, E. Revilla, and A. B. Smith. 2012. Approaching a state shift in Earth's biosphere. *Nature* 486: 52–58.

Beever, E. A. 2006. Monitoring biological diversity: strategies, tools, limitations, and challenges. *Northwestern Naturalist* 87: 66–79.

Beever, E. A., and J. E. Herrick. 2006. Effects of feral horses in Great Basin landscapes on soils and ants: Direct and indirect mechanisms. *Journal of Arid Environments* 66: 96–112.

Beever, E. A., L. E. Hall, J. Varner, A. E. Loosen, J. B. Dunham, M. K. Gahl, F. A. Smith, and J. J. Lawler. 2017. Behavioral flexibility as a mechanism for coping with climate change. *Frontiers in Ecology and the Environment* 15: 299–308.

Beever, E. A., L. Huntsinger, and S. L. Petersen. 2018. Conservation challenges emerging from free-roaming horse management: a vexing social-ecological mismatch. *Biological Conservation* 226: 321–328.

Beever, E. A., M. Huso, and D. A. Pyke. 2006. Multiscale responses of soil stability and invasive plants to removal of non-native grazers from an arid conservation reserve. *Diversity and Distributions* 12: 258–268.

Beever, E. A., R. J. Tausch, and W. E. Thogmartin. 2008. Multi-scale responses of vegetation to removal of horse grazing from Great Basin (USA) mountain ranges. *Plant Ecology* 196: 163–184.

Benedetti-Cecchi, L. I. Bertocci, and E. Maggi. 2006. Temporal variance reverses the impact of high mean intensity of stress in climate change experiments. *Ecology* 87: 2489–2499.

Bestelmeyer, B. T., A. Ash, J. R. Brown, B. Densambuu, M. Fernandez-Gimenez, J. Johanson, M. Levi, D. Lopez, R. Peinetti, L. Rumpff, and P. Shaver. 2017. State and transition models: theory, applications, and challenges. In D. D. Briske, editor. *Rangeland Systems: Processes, Management, and Challenges*, 303–346. Cham, Switzerland: SpringerOpen.

Bestelmeyer, B. T., J. R. Miller, and J. A. Wiens. 2003a. Applying species diversity theory to land management. *Ecological Applications* 13: 1750–1761.

Bestelmeyer, B. T., D. P. Goolsby, and S. R. Archer. 2011. Spatial perspectives in state-and-transition models: a missing link to land management? *Journal of Applied Ecology* 48: 746–757.

Bestelmeyer, B. T., J. R. Brown, K. M. Havstad, R. Alexander, G. Chavez, and J. E. Herrick. 2003b. Development and use of state-and-transition models for rangelands. *Journal of Rangeland Management* 56: 114–126.

Bisigato, A. J., M. B. Bertiller, J. O. Ares, G. E. Pazos, and F. Pugnaire. 2005. Effect of grazing on plant patterns in arid ecosystems of Patagonian monte. *Ecography* 28: 561–572.

Bisigato, A. J., R. M. L. Laphitz, and A. L. Carrera. 2008. Non-linear relationships between grazing pressure and conservation of soil resources in Patagonian Monte shrublands. *Journal of Arid Environments* 72: 1464–1475.

Bond, N. R., N. Grigg, J. Roberts, H. McGinness, D. Nielsen, M. O'Brien, I. Overton, C. Pollino, J. R. W. Reid, and D. Stratford. 2019. Assessment of environmental flow scenarios using state-and-transition models. *Freshwater Biology* 63: 804–816.

Boyce, M. S. 2006. Scale for resource selection functions. *Diversity and Distribution* 12: 269–276.

Brady, M. J., and N. A. Slade. 2004. Long-term dynamics of a grassland rodent community. *Journal of Mammalogy* 85: 552–561.

Briske, D. D., S. D. Fuhlendorf, and F. E. Smeins. 2005. State-and-transition models, thresholds, and rangeland health: a synthesis of ecological concepts and perspectives. *Rangeland Ecology and Management* 58: 1–10.

Brost, B. M., and P. Beier. 2012. Use of land facets to design linkages for climate change. *Ecological Applications* 22: 87–103.

Brown, J. H., 1995. *Macroecology.* Chicago, IL: University of Chicago Press.

Burke, M., K. Jorde, and J. M. Buffington. 2009. Application of a hierarchical framework for assessing environmental impacts of dam operation: changes in streamflow, bed mobility and recruitment of riparian trees in a western North American river. *Journal of Environmental Management* 90: S224–S236.

Calvert, J., C. McGonigle, S. A. Sethi, B. Harris, R. Quinn, and J. Grabowski. 2018. Dynamic occupancy modeling of temperate marine fish in area-based closures. *Ecology and Evolution* 8: 10192–10205.

Caro, T. 2010. *Conservation by Proxy: Indicator, Umbrella, Keystone, Flagship, and Other Surrogate Species.* Covelo, CA: Island Press.

Carreon-Martinez, L. B., T. B. Johnson, S. A. Ludsin, and D. D. Heath. 2011. Utilization of stomach content DNA to determine diet diversity in piscivorous fishes. *Journal of Fish Biology* 78: 1170–1182.

Chamberlain, S. A., J. L. Bronstein, and J. A. Rudgers. 2014. How context dependent are species interactions? *Ecology Letters* 17: 881–890.

Clements, F. E. 1916. *Plant Succession: An Analysis of the Development of Vegetation.* Washington, DC: Carnegie Institution of Washington.

Collins, S. L., S. R. Carpenter, S. M. Swinton, D. E. Orenstein, D. L. Childers, T. L. Gragson, N. B. Grimm, J. M. Grove, S. L. Harlan, J. P. Kaye, A. K. Knapp, G. P. Kofinas, J. J. Magnuson, W. H. McDowell, J. M. Melack, L. A. Ogden, G. P. Robertson, M. D. Smith, and A. C. Whitmer. 2011. An integrated conceptual framework for long-term social-ecological research. *Frontier in Ecology and the Environment* 9: 351–357.

Colwell, R. K., and D. C. Lees. 2000. The mid-domain effect: geometric constraints on the geography of species richness. *Trends in Ecology & Evolution* 15: 70–76.

Connell, J. H. 1978. Diversity in tropical rain forests and coral reefs. *Science* 199: 1302–1310.

Connell, J. H., and R. O. Slatyer. 1977. Mechanisms of succession in natural communities and their role in community stability and organization. *The American Naturalist* 111: 1119–1144.

Coops, N. C., S. N. Gillanders, M. A. Wulder, S. E. Gergel, T. Nelson, and N. R. Goodwin. 2010. Assessing changes in forest fragmentation following infestation using time series Landsat imagery. *Forest Ecology and Management* 259: 2355–2365.

Coumou, D., and S. Rahmstorf. 2012. A decade of weather extremes. *Nature Climate Change* 2: 491–496.

Cracraft, J. 1985. Biological diversification and its causes. *Annals of the Missouri Botanical Gardens* 72: 794–822.

Cutting, K. A., M. L. Anderson, E. A. Beever, S. R. Schroff, N. Korb, and S. McWilliams. 2016. Niche shifts and energetic condition of songbirds in response to phenology of food-resource availability in a high-elevation sagebrush ecosystem. *The Auk* 133: 685–697.

Dakos, V., B. Matthews, A. P. Hendry, J. Levine, N. Loeuille, J. Norberg, P. Nosil, M. Scheffer, and L. DeMeester. 2019. Ecosystem tipping points in an evolving world. *Nature Ecology and Evolution* 3: 355–362.

Dakos, V., S. R. Carpenter, W. A. Brock, A. M. Ellison, V. Guttal, A. R. Ives, S. Kéfi, V. Livina, D. A. Seekell, E. H. van Nes, and M. Scheffer. 2012. Methods for detecting early warnings of critical transitions in time series illustrated using simulated ecological data. *PLoS ONE* 7: e41010.

Dale, V. H., A. E. Lugo, J. A. MacMahon, and S. T. A. Pickett. 1998. Ecosystem management in the context of large, infrequent disturbances. *Ecosystems* 1: 546–557.

Daniel, C. J., L. Frid, B. M., Sleeter, and M.-J. Fortin. 2016. State-and-transition simulation models: a framework for forecasting landscape change. *Methods in Ecology and Evolution* 7: 1413–1423.

Darwin, C. 1859. *On the Origin of Species.* Reprinted 1964. Cambridge, MA: Harvard University Press.

de Valpine, P. 2002. Review of methods for fitting time-series models with process and observation error and likelihood calculations for nonlinear, non-Gaussian state-space models. *Bulletin of Marine Science* 70: 455–471.

DellaSala, D. A., and C. T. Hanson. 2015. *The Ecological Importance of Mixed-Severity Fires: Nature's Phoenix.* San Francisco, CA: Elsevier.

DeLong, D. C. 1996. Defining biodiversity. *Wildlife Society Bulletin* 24: 738–749.

Desoyza, A. G., W. G. Whitford, S. J. Turner, J. W. Van Zee, and A. R. Johnson. 2000. Assessing and monitoring the health of western rangeland watersheds. In S. S. Sandhu, B. D. Melzian, E. R. Long, W. G. Whitford, and B. T. Walton, editors. *Monitoring Ecological Condition in the Western United States: Proceedings of the Fourth Symposium on the Environmental Monitoring and Assessment Program (EMAP),* 153–166. San Francisco, CA, April 6–8, 1999. Dordrecht: Springer Netherlands.

Diaz, S., S. Lavorel, S. McIntyre, V. Falczuk, F. Casanoves, D. G. Milchunas, C. Skarpe, G. Rusch, M. Sternberg, I. Noy-Meir, J. Landsberg, W. Zhang, H. Clark, and B. D. Campbell. 2007. Plant trait responses to grazing – a global synthesis. *Global Change Biology* 13: 313–341.

Dobrowski, S. Z. 2011. A climatic basis for microrefugia: the influence of terrain on climate. *Global Change Biology* 17: 1022–1035.

Donohue, I., H. Hillebrand, J. M. Montoya, O. L. Petchey, S. L. Pimm, M. S. Fowler, K. Healy, A. L. Jackson, M. Lurgi, D. McClean, N. E. O'Connor, E. J. O'Gorman, and Q. Yang. 2016. Navigating the complexity of ecological stability. *Ecology Letters* 19: 1172–1185.

Early, R., B. A. Bradley, J. S. Dukes, J. J. Lawler, J. D. Olden, D. M. Blumenthal, P. Gonzalez, E. D. Grosholz, I. Ibanez, L. P. Miller, C. J. B. Sorte, and A. J. Tatem. 2015. Global threats from invasive alien species in the twenty-first century and national response capacities. *Nature Communications* 7: 12485.

Egler, F. E. 1954. Vegetation science concepts I. Initial floristic composition, a factor in oldfield vegetation development with 2 figs. *Plant Ecology* 4: 412–417.

Evans, N. T., B. P. Olds, M. A. Renshaw, C. R. Turner, Y. Li, C. L. Jerde, A. R. Mahon, M. E. Pfrender, G. A. Lamberti, and D. Lodge. 2016. Quantification of mesocosm fish and amphibian species diversity via environmental DNA metabarcoding. *Molecular Ecology Resources* 16: 29–41.

Fine, P. V. A. 2015. Ecological and evolutionary drivers of geographic variation in species diversity. *Annual Review of Ecology, Evolution, and Systematics* 46: 369–392.

Fisher, R. A. 1930. *The Genetical Theory of Natural Selection*. Oxford, UK: Clarendon Press.

Fisher, R. A. 1958. *The Genetical Theory of Natural Selection*. New York City: Dover.

Foley, M. M., R. G. Martone, M. D. Fox, C. V. Kappel, L. A. Mease, A. L. Erickson, B. S. Halpern, K. A. Selkoe, P. Taylor, and C. Scarborough. 2015. Using ecological thresholds to inform resource management: current options and future possibilities. *Frontiers in Marine Science* 2: 95.

Folke, C. J. 2016. Resilience (republished). *Ecology and Society* 21: 44.

Forester, J. D., A. R. Ives, M. G. Turner, D. P. Anderson, D. Fortin, H. L. Beyer, D. W. Smith, and M. S. Boyce. 2007. State–space models link elk movement patterns to landscape characteristics in Yellowstone National Park. *Ecological Monographs* 77: 285–299.

Fortin, M.-J., and M. R. T. Dale. 2009. *Spatial Analysis: A Guide for Ecologists*. New York: Cambridge University Press.

Fowler, N. L. 2002. The joint effects of grazing, competition, and topographic position on six savanna grasses. *Ecology* 83: 2477–2488.

Fox, B. J. 1982. Fire and mammalian secondary succession in an Australian coastal heath. *Ecology* 63: 1332–1341.

Fox, J. W. 2013. The intermediate disturbance hypothesis should be abandoned. *Trends in Ecology & Evolution* 28: 86–92.

Franklin, J. F., D. Lindenmayer, J. A. MacMahon, A. McKee, J. Magnuson, D. A. Perry, R. Waide, and D. Foster. 2000. Threads of continuity. *Conservation in Practice* 1: 8–17.

Franklin, J. F., K. Cromack, W. Denison, A. McKee, C. Maser, F. Swanson, and G. Juday. 1981. *Ecological Characteristics of Old-Growth Douglas-Fir Forests*. General Technical Report PNW-GTR-118. Portland, OR: U.S. Department of Agriculture, Forest Service, Pacific Northwest Forest and Range Experiment Station. 48 pp.

Gerstner, K., C. F. Dormann, A. Stein, A. M. Manceur, and R. Seppelt. 2014. Effects of land use on plant diversity – a global meta-analysis. *Journal of Applied Ecology* 51: 1690–1700.

González-Moreno, P., J. M. Diez, I. Ibáñez, X. Font, and M. Vilà. 2014. Plant invasions are context-dependent: multiscale effects of climate, human activity and habitat. *Diversity and Distributions* 20: 720–731.

Govindan, B. N., M. Kéry, and R. K. Swihart. 2011. Host selection and responses to forest fragmentation in acorn weevils: inferences from dynamic occupancy models. *Oikos* 121: 623–633.

Greenville, A. C., G. M. Wardle, V. Nyugen, and C. R. Dickman. 2016. Population dynamics of desert mammals: similarities and contrasts within a multi-species assemblage. *Ecosphere* 7: e01343.

Hällfors, M. H., J. Liao, J. Dzurisin, R. Grundel, M. Hyvarinen, K. Towle, G. C. Wu, and J. J. Hellmann. 2016. Addressing potential local adaptation in species distribution models: implications for conservation under climate change. *Ecological Applications* 26: 1154–1169.

Hargreaves, A. L., and C. G. Eckert. 2019. Local adaptation primes cold-edge populations for range expansion but not warming-induced range shifts. *Ecology Letters* 22: 78–88.

Harrison, S., B. D. Inouye, and H. D. Safford. 2003. Ecological heterogeneity in the effects of grazing and fire on grassland diversity. *Conservation Biology* 17: 837–845.

Hawkes, C. V., and J. J. Sullivan. 2001. The impact of herbivory on plants in different resource conditions: a meta-analysis. *Ecology* 82: 2045–2058.

Heilpern, S. A., B. Weeks, and S. Naeem. 2018. Predicting ecosystem vulnerability to biodiversity loss from community composition. *Ecology* 99: 1099–1107.

Hengeveld, R., and J. Haeck. 1982. The distribution of abundance. I. Measurements. *Journal of Biogeography* 9: 303–316.

Hiebeler, D. E., and B. R. Morin. 2007. The effect of static and dynamic spatially structured disturbances on a locally dispersing population. *Journal of Theoretical Biology* 246: 136–144.

Hobbs, R. J. 2001. Synergisms among habitat fragmentation, livestock grazing, and biotic invasions in southwestern Australia. *Conservation Biology* 15: 1522–1528.

Hobbs, R. J., E. Higgs, and J. A. Harris. 2009. Novel ecosystems: implications for conservation and restoration. *Trends in Ecology and Evolution* 24: 599–605.

Holling, C. S. 1973. Resilience and stability of ecological systems. *Annual Review of Ecology and Systematics* 4: 1–23.

Holmes, E. E., E. J. Ward, and K. Wills. 2012. MARSS: Multivariate autoregressive state-space models for analyzing time-series data. *R Journal* 4: 11–19.

Hurteau, M., H. Zald, and M. North. 2007. Species-specific response to climate reconstruction in upper-elevation mixed-conifer forests of the western Sierra Nevada, California. *Canadian Journal of Forest Research* 37: 1681–1691.

Ikeda, D. H., T. L. Max, G. J. Allan, M. K. Lau, S. M. Shuster, and T. G. Whitham. 2017. Genetically informed ecological niche models improve climate change predictions. *Global Change Biology* 23: 164–176.

Jakimow, B., P. Griffiths, S. van der Linden, and P. Hostert. 2018. Mapping pasture management in the Brazilian Amazon from dense Landsat time series. *Remote Sensing of Environment* 205: 453–468.

Jones, A. 2000. Effects of cattle grazing on North American arid ecosystems: a quantitative review. *Western North American Naturalist* 60: 155–164.

Karr, J. R. 1981. Assessment of biotic integrity using fish communities. *Fisheries* 6: 21–27.

Kéry, M., and J. A. Royle. 2015. *Applied Hierarchical Modeling in Ecology*. Boston: Academic Press.

Kéry, M., and M. Schaub. 2012. *Bayesian Population Analysis using WinBUGS: A Hierarchical Perspective*. San Diego, CA: Elsevier.

Killick, R., and I. A. Eckley. 2014. Changepoint: an R package for changepoint analysis. *Journal of Statistical Software* 58: 1–19.

Koerner, S. E., M. D. Smith, D. E. Burkepile, N. P. Hanan, M. L. Avolio, S. L. Collins, A. K. Knapp, N. P. Lemoine, E. J. Forrestel, S. Eby, D. I. Thompson, G. A. Aguado-Santacruz, J. P. Anderson, T. M. Anderson, A. Angassa, S. Bagchi, E. S. Bakker, G. Bastin, L. E. Baur, K. H. Beard, E. A. Beever, P. J. Bohlen, E. H. Boughton, D. Canestro, A. Cesa, E. Chaneton, J. Cheng, C. M. D'Antonio, C. Deleglise, F. Dembélé, J. Dorrough, D. J. Eldridge, B. Fernandez-Going, S. Fernández-Lugo, L. H. Fraser, B. Freedman, G. García-Salgado, J. R. Goheen, L. Guo, S. Husheer, M. Karembé, J. M. H. Knops, T. Kraaij, A. Kulmatiski, M.-M. Kytöviita, F. Lezama, G. Loucougaray, A. Loydi, D. G. Milchunas, S. J. Milton, J. W. Morgan, C. Moxham, K. C. Nehring, H. Olff, T. M. Palmer, S. Rebollo, C. Riginos, A. C. Risch, M. Rueda, M. Sankaran, T. Sasaki, K. A. Schoenecker, N. L. Schultz, M. Schütz, A. Schwabe, F. Siebert, C. Smit, K. A. Stahlheber, C. Storm, D. J. Strong, J. Su, Y. V. Tiruvaimozhi, C. Tyler, J. Val, M. L. Vandegehuchte, K. E. Veblen, L. T. Vermeire, D. Ward, J. Wu, T. P. Young, Q. Yu, and T. J. Zelikova. 2018. Change in dominance determines herbivore effects on plant biodiversity. *Nature Ecology & Evolution* 2: 1925–1932.

Lake, P. S. 2000. Disturbance, patchiness, and diversity in streams. *Journal of the North American Benthological Society* 19: 573–592.

Landres, P. B. 1992. Ecological indicators: panacea or liability? In: D. H. McKenzie, D. E. Hyatt, and V. I. McDonald, editors, *Ecological Indicators*, 1295–1318. Amsterdam, the Netherlands: Elsevier Applied Scientific Publishers.

Lawler, J. J., D. White, J. C. Sifneos, and L. L. Master. 2003. Rare species and the use of indicator groups for conservation planning. *Conservation Biology* 17: 875–882.

Leslie, P. and J. T. McCabe. 2013. Response diversity and resilience in social-ecological systems. *Current Anthropology* 54: 114–143.

Liao, J., Z. Ying, D. A. Woolnough, A. D. Miller, Z. Li, I. Nijs. 2016. Coexistence of species with different dispersal across landscapes: a critical role of spatial correlation in disturbance. *Proceedings of the Royal Society B: Biological Sciences* 283: 20160537.

Lindenmayer, D. B. 2009. *Forest Pattern and Ecological Process: A Synthesis of 25 Years of Research*. Melbourne, Australia: CSIRO Publishing.

Linder, H. L., J. K. Horne, and E. J. Ward. 2017. Modeling baseline conditions of ecological indicators: marine renewable energy environmental monitoring. *Ecological Indicators* 83: 178–191.

Litzow, M. A., F. J. Mueter, and J. D. Urban. 2013. Rising catch variability preceded historical fisheries collapses in Alaska. *Ecological Applications* 23: 1475–1487.

Loureau, M., and N. Mouquet. 1999. Immigration and the maintenance of local species diversity. *American Naturalist* 154: 427–440.

Lynch, A. J., L. M. Thompson, E. A. Beever, A. C. Engman, S. T. Jackson, T. J. Krabbenhoft, D. J. Lawrence, D. Limpinsel, R. T. Magill, T. A. Melvin, J. M. Morton, R. A. Newman, J. Peterson, M. T. Porath, F. J. Rahel, S. A. Sethi, and J. L. Wilkening. Guiding principles for managing ecosystem transformation. *Invited manuscript; revised manuscript in review. Journal of Wildlife Management.*

MacArthur, R. H., and E. O. Wilson. 1963. An equilibrium theory of insular zoogeography. *Evolution* 17: 373–387.

MacKenzie, D. I., J. D. Nichols, G. B. Lachman, S. Droege, J. A. Royle and C. A. Langtimm. 2002. Estimating site occupancy rates when detection probabilities are less than one. *Ecology* 83: 2248–2255.

Mackey, R. L., and D. J. Currie. 2001. The diversity disturbance relationship: is it generally strong and peaked? *Ecology* 82: 3479–3492.

Magurran, A. E. 2004. *Measuring Biological Diversity.* Malden, MA: Blackwell Publishing.

Manly, B. F. J., L. L. McDonald, D. L. Thomas, T. L. McDonald, and W. P. Erickson, editors. 2002. *Resource Selection by Animals: Statistical Design and Analysis for Field Studies,* 2nd edition. Dordrecht, the Netherlands: Kluwer Academic.

Marcot, B. G., M. A. Castellano, J. A. Christy, L. K. Croft, J. F. Lehmkuhl, R. H. Naney, K. Nelson, C. G. Niwa, R. E. Rosentreter, R. E. Sandquist, B. C. Wales, and E. Zieroth. 1999. Terrestrial ecology assessment. In T. M. Quigley, editor. *The Interior Columbia Basin Ecosystem Management Project: Scientific Assessment,* 1497–1713. Portland, OR: USDA Forest Service, Pacific Northwest Research Station. General Technical Report PNW-GTR-405.

Martin, J., V. Tolon, N. Morellet, H. Santin-Janin, A. Licoppe, C. Fischer, J. Bombois, P. Patthey, E. Pesenti, D. Chenesseau, and S. Said. 2018. Common drivers of seasonal movements on the migration – residency behavior continuum in a large herbivore. *Scientific Reports* 8: 7631.

May, R. M., S. A. Levin, and G. Sugihara. 2008. Ecology for bankers. *Nature* 451: 893–895.

Mayr, E. 1963. *Animal Species and Evolution.* Cambridge, MA: Harvard University Press.

McCune, B., and J. B. Grace. 2002. *Analysis of Ecological Communities.* Gleneden Beach, OR: MjM Software Design.

Milchunas, D. G., and I. Noy-Meir. 2002. Grazing refuges, external avoidance of herbivory and plant diversity. *Oikos* 99: 113–130.

Milchunas, D. G., W. K. Lauenroth, and I. C. Burke. 1998. Livestock grazing: animal and plant biodiversity of shortgrass steppe and the relationship to ecosystem function. *Oikos* 83: 65–74.

Moran, P. A. P. 1950. Notes on continuous stochastic phenomena. *Biometrika* 37: 17–23.

Mori, A. S. 2011. Ecosystem management based on natural disturbances: hierarchical context and non-equilibrium paradigm. *Journal of Applied Ecology* 48: 280–292.

Neubert, M. G., and H. Caswell. 1997. Alternatives to resilience for measuring the responses of ecological-systems to perturbations. *Ecology* 78: 653–665.

Niemi, G. J., and M. E. McDonald. 2004. Application of ecological indicators. *Annual Review of Ecology Evolution and Systematics* 35: 89–111.

Nolan, C., J. T. Overpeck, J. R. M. Allen, P. M. Anderso, J. L. Betancourt, H. A. Binney, S. Brewer, M. B. Bush, B. M. Chase, R. Cheddadi, M. Djamali, J. Dodson, M. E. Edwards, W. D. Gosling, S. Haberle, S. C. Hotchkiss, B. Huntley, S. J. Ivory, A. P. Kershaw, S.-H. Kim, C. Latorre, M. Leydet, A.-M. Lézine, K.-B. Liu, Y. Liu, A. V. Lozhkin, M. S. McGlone, R. A. Marchant, A. Momohara, P. I. Moreno, S. Müller, B. L. Otto-Bliesner, C. Shen, J. Stevenson, H. Takahara, P. E. Tarasov, J. Tipton, A. Vincens, C. Weng, Q. Xu, Z. Zheng, and S. T. Jackson. 2018. Past and future global transformation of terrestrial ecosystems under climate change. *Science* 361: 920–923.

Noss, R. F. 1990. Indicators for monitoring biodiversity: a hierarchical approach. *Conservation Biology* 4: 355–364.

Olden, J. D., J. J. Lawler, and N. L. Poff. 2008. Machine learning methods without tears: a primer for ecologists. *Quarterly Review of Biology* 83: 171–193.

Olson, D. M., and E. Dinerstein, 2002. The Global 200: Priority ecoregions for global conservation. *Annals of the Missouri Botanical Garden* 89(2): 199–224.

Osenberg, C. W., O. Sarnelle, S. D. Cooper, and R. D. Holt. 1999. Resolving ecological questions through meta-analysis: goals, metrics, and models. *Ecology* 80: 1105–1117.

Paine, R. T. 1966. Food web complexity and species diversity. *American Naturalist* 100: 65–75.

Pedersen, M. W., C. W. Berg, U. H. Thygesen, A. Nielsen, and H. Madsen. 2011. Estimation methods for nonlinear state-space models in ecology. *Ecological Modelling* 222: 1394–1400.

Perry, D. A., and M. P. Amaranthus. 1997. Disturbance, recovery, and stability. In K. A. Hohm and J. F. Franklin, editors. *Creating a Forestry for the 21st Century: The Science of Ecosystem Management*, 31–56. Washington, DC: Island Press.

Peters, D. P. C., O. E. Sala, C. D. Allen, A. Covich, and M. Brunson. 2007. Cascading events in linked ecological and socioeconomic systems. *Frontiers in Ecology and the Environment* 5: 221–224.

Peters, D. P. C., R. A. Pielke, B. T. Bestelmeyer, C. D. Allen, S. Munson-McGee, and K. M. Havstad. 2004. Cross-scale interactions, nonlinearities, and forecasting catastrophic events. *Proceedings of the National Academy of Sciences of the United States of America* 101: 15130–15135.

Pianka, E. R. 1994. *Evolutionary Ecology*. New York: Harper-Collins College Publishers.

Pulsford, S. A., D. B. Lindenmayer, and D. A. Driscoll. 2016. A succession of theories: purging redundancy from disturbance theory. *Biological Reviews* 91: 148–167.

Rahmstorf, S., and D. Coumou. 2011. Increase of extreme events in a warming world. *Proceedings of the National Academy of Sciences of the United States of America* 108: 17905–17909.

Reeves, J., J. Chen, X. L. Wang, R. Lund, and Q. Lu. 2007. A review and comparison of changepoint detection techniques for climate data. *Journal of Applied Meteorology and Climatology* 46: 900–915.

Roberge, J. M., and P. Angelstam. 2004. Usefulness of the umbrella species concept as a conservation tool. *Conservation Biology* 18: 76–85.

Rosenzweig, M. L. 1995. *Species Diversity in Space and Time*. Cambridge, UK: Cambridge University Press.

Roxburgh, S. H., K. Shea, and B. Wilson. 2004. The intermediate disturbance hypothesis: Patch dynamics and mechanisms of species coexistence. *Ecology* 85: 359–371.

Royle, J. A., and M. Kéry. 2007. A Bayesian state-space formulation of dynamic occupancy models. *Ecology* 88: 1813–1823.

Royle, J. A., and R. M. Dorazio. 2008. *Hierarchical Modeling and Inference in Ecology*. Boston, MA: Academic Press.

Sabo, J. L., and Post, D. M. 2008. Quantifying periodic, stochastic, and catastrophic environmental variation. *Ecological Monographs* 78: 19–40.

Sagarin, R. D., and S. D. Gaines. 2002. The 'abundant centre' distribution: to what extent is it a biogeographical rule? *Ecology Letters* 5: 137–147.

Sagarin, R. D., S. D. Gaines, and B. Gaylord. 2006. Moving beyond assumptions to understand abundance distributions across the ranges of species. *Trends in Ecology & Evolution* 21: 524–530.

Samis, K. E., and C. R. G. Eckert. 2007. Testing the abundant center model using range-wide demographic surveys of two coastal dune plants. *Ecology* 88: 1747–1758.

Scheffer, M., J. Bascompte, W. A. Brock, V. Brovkin, S. R. Carpenter, V. Dakos, H. Held, E. H. van Nes, M. Rietkerk, and G. Sugihara. 2009. Early-warning signals for critical transitions. *Nature* 46: 53–59.

Scheffer, M., S. Carpenter, J. A. Foley, C. Folke, and B. Walker. 2001. Catastrophic shifts in ecosystems. *Nature* 41: 591–596.

Schwartz, C. F., E. C. Thor, and G. H. Elsner. 1976. *Wildland Planning Glossary*. Berkeley, CA, USA: USDA Forest Service, Pacific Southwest Forest and Range Experimental Station. General Technical Report PSW-GTR-13. 252 p.

Scott, M. L., G. T. Auble, and J. M Friedman. 1997. Flood dependency of cottonwood establishment along the Missouri River, Montana, USA. *Ecological Applications* 7: 677–690.

Scott, J. M., E. D. Ables, T. C. Edwards, R. L. Eng, T. A. Gavin, L. D. Harris, J. B. Haufler, W. M. Healy, F. L. Knopf, O. Torgerson, and H. P. Weeks. 1995. Conservation of biological diversity: perspectives and the future for the wildlife profession. *Wildlife Society Bulletin* 23: 646–657.

Sethi, S. A., W. Larson, K. Turnquist, and D. Isermann. 2019. Estimating the number of contributors to DNA mixtures provides a novel tool for ecology. *Methods in Ecology and Evolution* 10: 109–119.

Sethi, S. A., M. Reimer, and G. Knapp. 2014. Alaskan fishing communities and the stabilizing role of fishing portfolios. *Marine Policy* 48: 134–141.

Shears, N. T., R. C. Babcock, and A. K. Salomon. 2008. Context-dependent effects of fishing: variation in trophic cascades across environmental gradients. *Ecological Applications* 18: 1860–1873.

Sih, A. 2013. Understanding variation in behavioural responses to human-induced rapid environmental change: a conceptual overview. *Animal Behaviour* 85: 1077–1088.

Sousa, W. P. 1984. The role of disturbance in natural communities. *Annual Review of Ecology and Systematics* 15: 353–391.

Spellerberg, I. F., and S. R. Hardes. 1992. *Biological Conservation*. Cambridge, UK: Cambridge University Press.

Stanley, E. H., S. M. Powers, and N. R. Lottig. 2010. The evolving legacy of disturbance in stream ecology: concepts, contributions, and coming challenges. *Journal of the North American Benthological Society* 29: 67–83.

Steffen, W., W. Broadgate, L. Deutsch, O. Gaffney, and C. Ludwig. 2015. The trajectory of the Anthropocene: the great acceleration. *The Anthropocene Review* 2: 81–98.

Sugihara, G., R. May, H. Ye, C. Hsieh, E. Deyle, M. Fogarty, and S. Munch. 2012. Detecting causality in complex ecosystems. *Science* 338: 496–500.

Teague, W. R., S. L. Dowhower, and J. A. Waggoner. 2004. Drought and grazing patch dynamics under different grazing management. *Journal of Arid Environments* 58: 97–117.

ter Braak, C. J. F. 1994. Canonical community ordination. Part I: basic theory and linear methods. *Ecoscience* 1: 127–140.

Thom, D., and R. Seidl. 2015. Natural disturbance impacts on ecosystem services and biodiversity in temperate and boreal forests. *Biological Reviews* 91: 760–781.

Thompson, L. M., A. J. Lynch, E. A. Beever, A. C. Engman, J. A. Falke, S. T. Jackson, T. J. Krabbenhoft, D. J. Lawrence, D. Limpinsel, R. T. Magill, T. A. Melvin, J. M. Morton, R. A. Newman, J. Peterson, M. T. Porath, F. J. Rahel, S. A. Sethi, and J. L. Wilkening. When is resistance futile? Resisting, accepting, or directing ecosystem transformation. *Invited manuscript; revised manuscript in review. Fisheries.*

Tilman, D., and J. A. Downing. 1994. Biodiversity and stability in grasslands. *Nature* 367: 363–365.

Varner, J., and M. D. Dearing. 2014. The importance of biologically relevant microclimates in habitat suitability assessments *PLoS ONE* 9: e104648.

Verbesselt, J., A. Zeileis, and M. Herold. 2012. Near real-time disturbance detection using satellite image time series. *Remote Sensing of Environment* 123: 98–108.

Vesk, P. A., M. R. Leishman, and M. Westoby. 2004. Simple traits do not predict grazing response in Australian dry shrublands and woodlands. *Journal of Applied Ecology* 41: 22–31.

Vilà, M., J. L. Espinar, J. Hejda, P. E. Hulme, V. Jarošík., J. L. Maron, J. Pergl, U. Schaffner, Y. Sun, and P. Pyšek. 2011. Ecological impacts of invasive alien plants: a meta-analysis of their effects on species, communities and ecosystems. *Ecology Letters* 14: 702–708.

Waples, R. S. 1995. Evolutionarily significant units and the conservation of biological diversity under the Endangered Species Act. *American Fisheries Society Symposium* 17: 8–27.

Warner, P. J., and J. H. Cushman. 2002. Influence of herbivores on a perennial plant: variation with life history stage and herbivore species. *Oecologia* 132: 77–85.

Weber, N., N. Bouwes, M. M. Pollock, C. Volk, J. M Wheaton, G. Wathen, J. Wirtz, and C. E. Jordan. 2017. Alteration of stream temperature by natural and artificial beaver dams. *PLoS ONE* 12: e0176313.

Westoby, M., B. H. Walker, and I. Noy-Meir. 1989. Opportunistic management for rangelands not at equilibrium. *Journal of Range Management* 42: 266–274.

Whitmee, S., A. Haines, C. Beyrer, F. Boltz, A. G. Capon, B. F. De Souza Dias, A. Ezeh, H. Frumkin, P. Gong, P. Head, R. Horton, G. M. Mace, R. Marten, S. S. Myers, S. Nishtar, S. A. Osofsky, S. K. Pattanayak, M. J. Pongsiri, C. Romanelli, A. Soucat, J. Vega, and D. Yach. 2015. Safeguarding human health in the Anthropocene epoch: Report of the Rockefeller Foundation-Lancet Commission on planetary health. *The Lancet* 386: 1973–2028.

Whittaker, R. H. 1960. Vegetation of the Siskiyou Mountains, Oregon and California. *Ecological Monographs* 30: 279–338.

Wickham, J. D., J. Wu, and D. F. Bradford. 1997. A conceptual framework for selecting and analyzing stressor data to study species richness at large spatial scales. *Environmental Management* 21: 247–257.

Wiens, J. A. 1989a. Spatial scaling in ecology. *Functional Ecology* 3: 385–397.

Wiens, J. A. 1989b. *The Ecology of Bird Communities. Volume 1. Foundations and Patterns*. Cambridge, UK: Cambridge University Press.

Williams, J. W., and S. T. Jackson. 2007. Novel climates, no-analog communities, and ecological surprises. *Frontiers in Ecology and the Environment* 5: 475–482.

Yu, Z., H. Wang, T. Wang, W. Sun, X. Yang, and J. Liu. 2015. Tick-borne pathogens and the vector potential of ticks in China. *Parasites & Vectors* 8: 24.

Zipper, C. E., J. A. Burger, J. G. Skousen, P. N. Angel, C. D. Barton, V. Davis, and J. A. Franklin. 2011. Restoring forests and associated ecosystem services on Appalachian coal surface mines. *Environmental Management* 47: 751–765.

Zweig, C. L., and W. M. Kitchens. 2009. Multi-state succession in wetlands: a novel use of state and transition models. *Ecology* 90: 1900–1909.

Forest Disturbances

Scale-dependence, Natural vs. Anthropogenic Disturbance, Length of Recovery or Succession

chapter two

Wildlife Responses to Abiotic Conditions, Herbivory, and Management of Aspen Communities

Henry Campa III and Alexandra Locher

Introduction

Natural-resources researchers and managers face many issues related to the multiple uses of forest resources, wildlife-habitat relationships within forest ecosystems, and conservation of forest diversity at multiple spatial and temporal scales. Aspen (*Populus* spp.) communities are important throughout much of North America because of their ecological contribution to biodiversity, their economic importance to the pulp industry, and as a potential renewable energy resource. For instance, numerous wildlife species depend on aspen to meet their life requirements (Stelfox 1995). Ruffed grouse (*Bonasa umbellus*) use catkins from mature male aspen as their staple winter food source (Gullion and Svoboda 1972). White-tailed deer (*Odocoileus virginianus*) rely heavily on aspen leaves and new shoots for spring and summer foraging (Kohn and Mooty 1971), and elk (*Cervus elaphus nelsoni*) in Michigan also use it for foraging and hiding cover (Beyer 1987). Parsons et al. (2003) found that bat species in British Columbia depend on the decay-causing fungus (*Phellinus tremulae*) in old aspen to create cavities for roosting. Pojar (1995) found bird communities, including neotropical migrants, to be associated with specific aspen seral stages, and Dingledine and Haufler (1983) found live aspen provided suitable habitat for cavity-nesting birds. Other wildlife such as woodcock (*Scolopax minor*), beaver (*Castor canadensis*), woodpeckers, and showshoe hares (*Lepus americanus*) also obtain food and cover requirements from aspen stands.

Aspen is also important economically. In the Great Lakes Region, aspen is the dominant species harvested for pulpwood, comprising 40 percent of the total timber harvested for paper, cardboard, particleboard, and other products (Piva 2007). In the past few decades, aspen has been studied as an important material for bioenergy, owing to its fast growth rate and ease of propagation (Dickmann 2006). In the western United States, aspen is less important economically, but is considered to be very biologically diverse and supports high avian species abundance and diversity (Worrall et al. 2008).

Because of the economic and ecological values of aspen communities, foresters, wildlife biologists, and natural resource managers have implemented various management and harvesting practices to intensify the productive capacity of aspen while hopefully maintaining the natural diversity of composition, structure, function, and processes (e.g., succession) occurring within aspen communities. However, biodiversity components within aspen communities may be at risk due to anthropogenic and/or natural disturbances that are either too frequent and/or too intense for aspen to sustain. Common disturbances in aspen communities include ungulate browsing (e.g., Raymer 2000), mortality from insect or fungal pathogens (Worrall et al. 2008), intensive timber harvesting practices (Locher et al. 2012), forest fire suppression, and climate change (Iverson and Prasad 1998). The occurrence of these disturbances, and outcomes of them, on aspen structure and composition may be unpredictable and, in some cases, cumulative, which in effect limit the ability of managers to forecast the nature and scope of the change that is likely to come.

Natural resources professionals realize that aspen communities are much more diverse compositionally and structurally than originally described due to the diversity of habitat types supporting aspen (Felix 2008; Rogers et al. 2014). A habitat type is an area capable of supporting similar seral communities (i.e., vegetation types) and successional pathways based on similar abiotic (e.g., soil texture, moisture, hydrology, topography) characteristics (Daubenmire 1966). Aspen is capable of growing in a variety of habitat types from those supported by poorly-drained silty soils to those supported by excessively-drained sands. In essence, not all aspen stands are ecologically equal. Based on this realization, to sustain the aspen resource across its range requires managers to adapt their management practices to (1) the ecological conditions in which aspen currently exists (antecedent conditions) and (2) the disturbance ecology in which aspen evolved.

Our objectives are to demonstrate (1) why aspen across an array of forest habitat types with different natural successional trajectories should be managed differently to conserve unique vegetation types that spatially and temporally contribute to wildlife conservation, and (2) how the ecological site potential of forest habitat types may influence the ability of managers to use forest management practices to mimic natural disturbances in aspen communities in the face of ungulate herbivory.

Disturbance Ecology of Aspen: Maintaining a Unique Natural Resource

Distribution and Characteristics of Aspen Communities at Multiple Scales

Aspen communities are the most widely distributed among all forest communities in North America (Perala 1990). Aspen grows trans-continentally at northern latitudes (>40°N) and higher altitudes (2100–3350 m; Perala 1990). They are characteristic of mountainous regions in western North America, and the northern deciduous forests in eastern North America. Aspen is considered an early-successional forest tree species. In areas where there are remnant individuals, aspen becomes well-established in even-aged stands following natural and anthropogenic disturbances such as fire, logging, extensive herbivory, or land-use practices exposing extensive areas of bare soil and open sunlight. Aspen evolved as pioneer species—that are ecologically well suited to colonize such disturbed sites (Graham et al. 1963) and can regenerate on exposed mineral soil by cottony seeds that may be dispersed several kilometers by the wind (Zasada and Phipps 2000). Additionally, aspen may regenerate in open areas through suckering, which is when vegetative shoots sprout from the horizontal roots of parental trees (Schier and Smith 1979). Suckers are produced when the tops of aspen trees are killed or removed by such disturbances as fire, wind, or logging.

The structure and composition of aspen stands, however, differs significantly between the western and eastern regions of North America and at smaller spatial scales within regions based on differences in climate, soil types, or other geologic factors. For example, in western North America, quaking aspen (*P. tremuloides*) dominates communities, but balsam popular (*P. balsamifera*) may occur in riparian areas of boreal and montane forests (Zasada and Phipps 1990; Figure 2.1). Quaking aspen communities in western regions mature between 80–100 years and can grow 30–100 m tall depending on site conditions and genetics (Mueggler 1985). Western aspen communities are most frequently even-aged and become established within 2–4 years from suckering after fire or another major disturbance such as clearcutting (Shepperd 1981; Jones and DeByle 1985). Intense browsing from livestock or wildlife species such as deer, elk, or beaver may also help maintain an even-aged structure (Johnston and Doty 1972; Campa 1989).

Uneven-aged aspen communities supporting trees with ages 10–20 years apart are also common throughout western North America (Davidson et al. 1959). These uneven-aged stands may be self-perpetuating, stable communities and may persist for centuries (Mueggler 1985). Aspen is not considered a predominant timber species in the west since it represents only 3 percent of timber volume harvested (Oswalt et al. 2014).

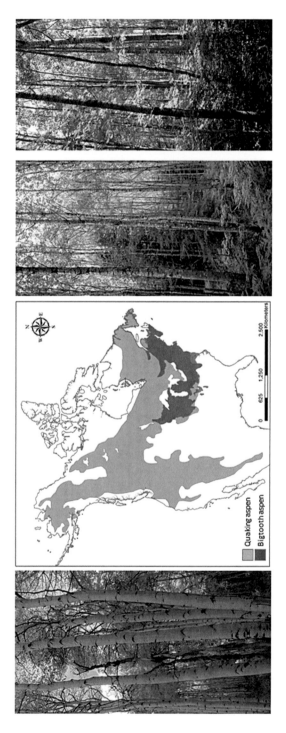

Figure 2.1 **(See color insert.)** The range of quaking and bigtooth aspen in North America. Throughout its range, there is tremendous diversity in aspen community composition and structure depending on climate, soils, and geology. Photos illustrate differences in aspen community structure in Colorado (left) and on xeric sandy and mesic loamy soils in Michigan (right). (Photo credits: (left) Pixabay-853565; (right) A. Felix.)

In eastern North America, aspen represents approximately 10 percent of timberland and 7 percent of growing stock volume (Oswalt et al. 2014). Most (85 percent) of the aspen harvested in the east comes from Minnesota, Michigan, and Wisconsin (Oswalt et al. 2014). Aspen communities in the east may be dominated by quaking aspen or bigtooth aspen (*P. grandidentata*; Figure 2.1). Balsam poplar may also occur occasionally on poorly drained soils as a component of communities dominated by lowland swamp conifers or lowland hardwoods (Zasada and Phipps 1990). Aspen communities in the east grow faster and mature sooner (i.e., 50–60 years old) than those in the west. Although some aspen trees may live to be over 100 years old, most communities over 70 years show signs of deterioration and senescence (Graham et al. 1963). Eastern aspen communities are frequently even-aged and may occur in essentially pure stands when they are maintained by frequent (e.g., 40–50 years; Felix 2008) even-aged forest management practices (e.g., clearcutting) to meet wildlife and/or timber management objectives. Depending on the soil conditions of eastern aspen communities, these even-aged stands may also consist of northern hardwood species (e.g., American beech—*Fagus grandifolia*, sugar maple—*Acer saccharum*) and conifer species (e.g., eastern white pine—*Pinus strobus*, balsam fir—*Abies balsamea*; Kotar and Burger 2000).

Currently, although the total area of forest land has decreased due to urbanization and land use, aspen still comprises approximately 25 percent of state-owned forests in Michigan (Holste and Garmon 2013). However, because aspen regenerates easily and is economically important to the timber industry and the economy of Michigan, it will likely maintain its relatively high representation as a vegetation community on forestland (Dickmann and Leefers 2004).

Aspen Disturbance Ecology

Prior to European settlement in the nineteenth century, aspen communities were much less prevalent in Michigan than they are today in the twenty-first century (e.g., Cleland et al. 2001). Natural disturbances perpetuating aspen communities, regardless of site, historically included frequent fire and large windfalls (Palik and Pregitzer 1992, Dickmann and Leefers 2004). Aspen, once considered a "weed species" (Spencer and Thorne 1972), became prominent within the Michigan landscape and the other Great Lakes States (Wisconsin, Minnesota) in the early part of the twentieth century. During this time period, aspen growth increased following intensive disturbances caused by frequent land clearing and logging of mature pine and hardwood forests and the subsequent fires that burned much of the state in the late 1800s and early 1900s (Haines and Sando 1969; Cleland et al. 2001; Dickmann and Leefers 2004). Pine and hardwoods were ultimately replaced with aspen as the resulting composition of Michigan's forests due to the disturbance regimes (e.g., timber harvesting, fire), physical environment (e.g., soil conditions), and biological processes (e.g., succession of shade-intolerant species; Cleland et al. 2001).

Cleland et al. (2001) reported that nearly 2.5 million hectares of forest-land in Michigan was burned between 1871 and 1908, and approximately 809,000 hectares burned between 1920 and 1929. The intensive natural disturbances, especially the intensive fires and frequent re-burns after timber harvests, created the optimal conditions for aspen to thrive. These conditions included vast areas of open sunlight, warm soil, the removal of parent trees for stimulating sucker production, and at this time low densities of large herbivores (e.g., white-tailed deer). Estimates from early General Land Office surveys that began in 1826 indicated that prior to these disturbances, less than 1 percent of the original forests in Michigan (approximately 15 million hectares of timberland) was composed of aspen (Stearns 1995). However, by the mid-1900s, aspen represented 25 percent of approximately eight million hectares of Michigan timberland (Findell et al. 1960).

In recent decades, disturbances from fire have been suppressed in landscapes. Today, frequent logging practices, and land use creating exposed soil are the primary planned disturbances perpetuating aspen (Dickmann and Leefers 2004). Even prescribed fire now plays an almost non-existent disturbance factor with regeneration of aspen (Cleland et al. 2001). Logging, typically clearcutting, removes overstory competition and releases aspen, which sprouts and grows rapidly from parent roots or by seeds that sprout in soil exposed from other disturbances, including windthrow, insect herbivory, and canopy openings created from disease.

Other disturbances that can damage or cause aspen mortality directly or indirectly include damage from insects such as the tent caterpillar (*Malacosoma americanum*), aspen tortix (*Choristoneura conflictana*), and poplar borer (*Saperda calcarata*); and infections from fungus such as the hypoxylon canker (Brinkman and Roe 1975), and sunscald (Hart et al. 1986). Studies suggest that aspen on poor-quality sites may be more susceptible to damage from such agents. Ungulate herbivory, if too frequent and/or intense, can also be a significant ecosystem disturbance factor if animals are locally abundant, affecting the physiognomy, function, composition, structure, and nutritional qualities of aspen communities (e.g., Campa 1989; Campa et al. 1992; McShea et al. 1997). Aspen is an important food source for many ungulate species including white-tailed deer, elk, and moose (*Alces alces*), and intense browsing pressure can shift the long-term composition and structure of aspen stands and, in the worst cases, impair regeneration or shift natural successional trajectories entirely (Campa 1989; Campa et al. 1992; Raymer 2000).

In the absence of large-scale, stand-replacing disturbances (natural or anthropogenic) creating even-aged stands, the dominant successional trajectories of aspen communities will change significantly. Frelich and Reich (1995) described that in the Boundary Waters Canoe Area Wilderness in Minnesota, fire-rotation periods of 50–100 years during presettlement times to more than 1000 years ago resulted in many stands being transformed to uneven-aged. For example, they reported that under historic disturbance patterns, fires would have maintained stands with canopies of jack pine (*Pinus banksiana*) or aspen. Under longer fire-return intervals, stands develop into later successional stages dominated by a mixture of species such as black spruce (*Picea mariana*) on wet soils, balsam fir and paper birch (*Betula papyrifera*) on dry soils, or northern hardwoods or hardwood-pine mixtures on mesic soils (Kotar and Burger 2000).

With the complex disturbance ecology of aspen, natural resource managers are challenged with trying to sustain the diversity of aspen communities with forest management practices that can emulate natural disturbances yet are sometimes poorly understood or accepted by some stakeholders (e.g., Clausen and Schroeder 2004). Additionally, little is known about the interactions among different disturbances (i.e., natural or anthropogenic) that may present an ecological threshold that disrupts natural mechanisms of recovery and regeneration following the disturbances. For instance, although aspen may recover from disturbance with which it evolved (e.g., fire), such a disturbance combined with increased extreme drought from climate change or increased browsing from high ungulate densities may prevent an aspen community from recovery (Frey et al. 2004).

Characteristics of Aspen in Michigan on Different Habitat Types

The distribution of aspen in Michigan and surrounding areas within the Great Lakes Region spans areas with poorly-drained mesic soils to those with well-drained xeric soils (Figure 2.2). Aspen communities found on different soil types tend to respond differently to disturbances, which leads to variable aspen forest structure and composition on varying substrates. For example, in the absence of natural or anthropogenic disturbances, an aspen community occurring on mesic sandy loams on ice contact ridges will have a successional trajectory leading to white pine and red maple (*Acer rubrum*) in late successional stages. Conversely, aspen occurring in early successional stages on loamy soils will progress to vegetation types dominated by sugar maple and black cherry (*Prunus serotina*) to sugar maple and eastern hemlock (*Tsuga canadensis*) in later stages (Felix 2008). These different structural and compositional characteristics among aspen communities are important for conserving components of biodiversity including a diversity of wildlife species and communities, and inherent ecological variation (Raymer 2000, Felix 2008). Many forest and wildlife managers are incorporating an understanding of the diverse structural and compositional differences among aspen communities occurring on different habitat types into plans to maintain the diversity in aspen communities over time (Felix 2008). For example, maintaining representation of aspen in a variety of age classes on the range of soil types that can support it is important in sustaining aspen diversity.

Bigtooth aspen is commonly found on dry, sandy soils (Burns and Honkala 1990). Common forest associates in the Great Lakes Region of the United States include jack pine, red pine (*P. resinosa*), white pine, northern pin oak (*Quercus ellipsoidalis*), northern red oak (*Q. rubra*), balsam fir, and red maple. Quaking aspen, however, can tolerate a wider range of conditions than can bigtooth aspen, and grows best on well-drained loamy soils and soils high in organic matter (Perala 1990). Common forest associates include balsam fir and white spruce (*Picea glauca*) on drier, coarser soils,

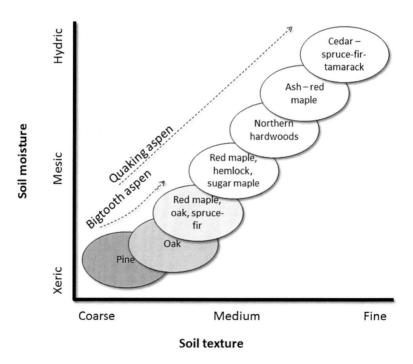

Figure 2.2 Forest types and soil characteristics associated with bigtooth aspen and quaking aspen communities in Michigan. Shaded ellipses represent ecosystems that support primarily bigtooth aspen. White ellipses represent ecosystems that support quaking aspen.

sugar maple and other northern hardwoods on mesic loams, and lowland conifers such as northern white cedar (*Thuja occidentalis*) on hydric soils.

Most (>60 percent) of the volume and area of aspen within Michigan is less than 20 years old, as indicated by the proportion of trees within the 2.54–7.6 cm (1–3 in) diameter class from 2014 Forest Inventory and Analysis data (Forest Inventory Data Online 2015). Approximately 26 percent of aspen is 20–40 years old, and 9 percent is 40–60 years old, which is the age range when aspen in Michigan is most economically valuable. Although aspen may live past 100 years old, most of the aspen in Michigan show signs of deterioration after 60 years old and lose their commercial value (Graham et al. 1963). As such, older age classes of aspen are least prevalent in Michigan landscapes because they are either harvested or begin to deteriorate and succumb to later successional stages. If aspen is not harvested or disturbed, shade-tolerant species (e.g., sugar maple) will eventually dominate the communities. Additionally, the distribution of aspen is not proportionally represented across all habitat types that can support aspen. For instance, habitat types characterized by mesic loamy soils support over 30 percent of the aspen occurring in upper Michigan, but cover less than 15 percent of the land area (Figure 2.3).

Ungulate Herbivory as a Disturbance on Aspen Communities across a Range of Ecological Conditions: Managing for Sustainability

Ungulate herbivory on plant communities is an ecosystem process that can dramatically influence the composition, structure, nutritional qualities, and natural successional trajectory of habitat types, especially if browsing or foraging pressures are too intense, frequent, and/or occur during the growing season (e.g., Campa, 1989; Campa et al. 1992; McShea 2012; Lovely et al. 2013). One of the contributing factors that has influenced these changes in forest ecosystems is that numbers of large herbivores, such as white-tailed deer, have become locally abundant in some areas across the Midwestern and Northeastern regions of the U.S. as disturbances to plant communities have created early successional vegetation types and edges (e.g., Dickmann and Leefers 2004; Campa et al. 2011). McCabe and McCabe (1997) provided a summary of the history of white-tailed deer numbers across the U.S. and point to the classic work of Ernest Seton who published articles in 1909 and 1929 citing that there was an estimated four white-tailed deer per km^2 across their range during what he refers to as "primitive times," or approximately 20 million deer (Seton 1909: 78). However, Seton (1929) states that white-tailed deer numbers could have increased to 40 million animals across the species' range before Europeans arrived. Upon further analysis of historical data,

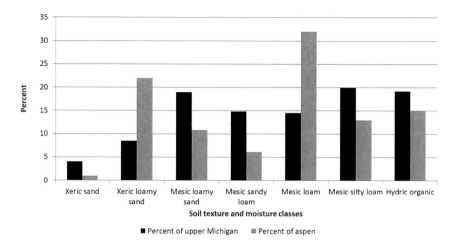

Figure 2.3 Comparison between percentage of Michigan's Upper Peninsula characterized by different soil texture and moisture classes, and percentage of aspen occupying those same soil texture and moisture classes. Data were compiled from literature and field data collected between 2004 and 2006.

McCabe and McCabe (1997), however, estimated an average density of white-tailed deer between 3.1 and 4.2 deer per km².

Today, white-tailed deer densities are substantially higher than historical estimates, in part due to the landscapes humans have created through forest management practices (e.g., clearcutting), rotations favoring early-successional vegetation, increased forest fragmentation, and land use change (i.e., agriculture and residential development). These landscape changes have produced suitable habitat components for the survival and reproduction of deer and in some cases provide refuges from harvests (e.g., Felix et al. 2004; Burroughs et al. 2006; Campa et al. 2011; Lovely et al. 2013; Hummel et al. 2018). Specifically, deer density estimates across the Midwestern region of the U.S. today can be greater than 10 deer per km² (e.g., Rooney and Waller 2003; Urbanek et al. 2012). For example, in northern lower Michigan (i.e., Alpena, Alcona, Montmorency, Oscoda Counties) the estimated deer density was 19–23 per km² in 1994 (Schmitt et al. 2002) prior to extensive anterless deer harvests to control the spread of bovine tuberculosis. Such a density of white-tailed deer above historic estimates now coupled with the existence of other herbivores, such as elk (e.g. in Michigan estimated between 834 and 1512 elk as of February 7, 2018, historically as low as 200) (MDNR 1975, 2018), now occurring in many eastern states in the U.S. (e.g., Michigan, Wisconsin, Pennsylvania, Kentucky) has complicated how forest ecosystems can be sustained to meet a diversity of ecological and social management objectives. What are the cumulative effects of multiple large herbivores on tree species and forest plant communities, especially when some plant species may not have evolved with the browsing and foraging pressures they face today and the fact that there is wide variability in the ability of plants to withstand today's browsing (e.g. Westell 1954; Campa et al. 1986)?

Although eastern elk (*Cervus elaphus canadensis*) historically occurred in Michigan until approximately 1877 (Murie 1951), Rocky Mountain elk were released in the early 1900s up to 1917 (Moran 1973) giving rise to a population that eventually was large enough to sustain at least one hunting season per year since 1984. In 1984, Michigan's elk herd within the Pigeon River Country and Atlanta State Forests was estimated to be 850 individuals, with the population growing at 16 percent per year (Beyer 1987). This growth in population was accomplished primarily through (1) decreased poaching as a result of law enforcement; and (2) improved habitat quality through more forest management, primarily clearcutting aspen and establishing wildlife openings (Moran 1973; Beyer 1987).

As elk responded to these management actions, natural resource managers became concerned about the browsing effects on bigtooth and quaking aspen, primarily by elk. Were browsing pressures so intense and frequent that ultimately browsing would decrease the availability of aspen for wood products and/or contribute to the loss of the aspen communities across Michigan's elk range?

Managing Aspen-elk and White-tailed Deer Herbivory in Northern Lower Michigan: A Case Study

Two factors contributing to northern Michigan's landscape being of high quality habitat for deer and elk are the availability and even-aged management of aspen and the fact that it can provide suitable food quality (e.g., Campa 1989; Campa et al. 1992). Today, natural resources managers are faced with a dilemma—the need to harvest aspen to meet numerous timber and wildlife management objectives, but by doing so are providing an abundance of regenerating aspen suckers for browsing and foraging, which may then help maintain relatively high densities of herbivores or at least make them locally abundant. The question remains how, when, and where can aspen be harvested to sustain it as a resource without contributing to browsing or foraging intensities so intense and frequent that it leads to aspen mortality and potential shifts in the availability of naturally occurring vegetation types for other wildlife species and communities?

In response to this forest-wildlife management challenge, Campa (1989) conducted a four-year study investigating the effects of deer and elk browsing on various early-aged bigtooth and quaking aspen clearcut communities' stand characteristics and nutritional qualities using replicated exclosures and paired areas open to browsing and foraging. Bigtooth aspen communities largely occurred on well-drained soils whereas quaking aspen occurred on a greater variety of mesic and poorly drained soils. Aspen community age classes that were 1–6 growing seasons old were also replicated throughout the study. Results indicated that the current annual growth of bigtooth aspen twigs was browsed (19–98 percent) significantly more than quaking aspen (5–75 percent) within each age class and this browsing pressure was causing significant changes in the stem densities and availability of cover of both species. For example, the density of trees >2.0 m in height in four-year-old bigtooth aspen exclosures (7400 stems/hectare) was significantly greater than in areas open to browsing (1950 stems/hectare (e.g., Figure 2.4; 1986 photos). A similar observation was made for the availability of cover in exclosures vs. areas open to browsing—that elk and deer browsing were decimating the mid- (0.5–2.0 m) and upper- (>2.0 m) height classes in these stands. Regenerating stands (e.g., six years old) of quaking aspen showed a similar response to browsing as bigtooth aspen on stem densities and cover; however, results between exclosures and areas open to browsing were not statistically different (e.g., Figure 2.5; 1986 photos). One explanation for these results was that bigtooth aspen was being browsed more heavily than quaking aspen due to its higher crude protein contents.

BTA #72; 4yr old; Exclosure 1986   BTA #72; 4yr old; Open to Browsing 1986

BTA #72; 17yr old; Exclosure 1999   BTA 17yr old; Open to Browsing 1999

Figure 2.4 (**See color insert.**) Stand structure of a 4-year-old bigtooth aspen (BTA; numbers correspond to stand location) community within a 20 × 20 m exclosure and a paired area open to white-tailed deer and elk browsing in northern lower Michigan in 1986. All exclosures were built on clearcuts immediately following clearcutting in 1982. The stand structure of the same aspen community within the exclosure and paired area 13 years later. (Photo credits: (top left and right) H. Campa III, 1986; (bottom left and right) D. Raymer, 1999.)

QA #14; 6yr old; Exclosure 1986   QA#14; 6yr old; Open to Browsing 1986

QA #14; 19yr old; Exclosure 1999   QA #14; 19yr old; Open to Browsing 1999

Figure 2.5 **(See color insert.)** Stand structure of a 6-year-old quaking aspen (QA; numbers correspond to stand location) community within a 20 × 20 m exclosure and a paired area open to white-tailed deer and elk browsing in northern lower Michigan in 1986. The stand structure of the same aspen community within the exclosure and paired area 13 years later. (Photo credits: (top left and right) H. Campa III, 1986; (bottom left and right) D. Raymer, 1999.)

Whereas the study above demonstrated how herbivory from large herbivores, likely above historic population sizes, can influence aspen communities, it was unknown what level or threshold of browsing aspen communities could sustain without being impacted either ecologically or economically. To investigate this question, Campa et al. (1992) conducted a study to investigate the effects of simulated ungulate browsing (0–100 percent clipping of current annual growth for years 1 or 2, or 1 and 2 years post-clearcutting) on aspen stand characteristics and nutritional qualities. This project was conducted within replicated exclosures for bigtooth and quaking aspen. Results indicated that when more than 50 percent of the current annual growth of bigtooth aspen was clipped to simulate browsing, tree heights were reduced by 24–32 percent and number of twigs reduced by 66–80 percent. Such results were a concern to natural resources professionals in this area given that browsing intensities on bigtooth aspen frequently exceeded 50 percent. Obviously, the long-term sustainability of aspen and the wildlife species and communities that depend on aspen for their habitat components would require an integrated approach of reducing elk and deer numbers while continuing to harvest aspen across the elk range to disperse browsing across multiple types of aspen communities.

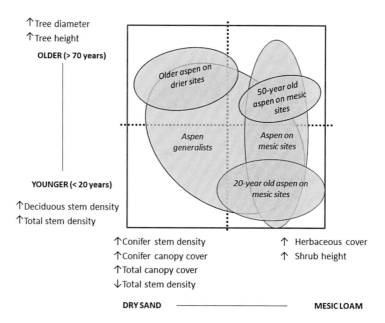

Figure 2.6 Conceptual diagram of avian community assemblages associated with various aspen age classes (i.e., following harvest) and habitat types (i.e., abiotic conditions) in northern Michigan. Assemblages were associated with different vegetation structural and compositional characteristics.

To better understand the long-term effects of herbivory as an ecosystem process and as a disturbance agent on aspen communities, Campa et al. (1993) and Raymer (2000) expanded on the work conducted in the 1980s using deer and elk exclosures paired with areas open to browsing. Campa documented that 10 years post-clearcutting, ungulates continued to browse bigtooth aspen more than quaking aspen and that these continued browsing pressures were continuing to impact stem densities, vertical and horizontal cover, and the frequency of some plant species. Raymer implemented a study to investigate (1) how reducing elk numbers (i.e. from 1350 in 1993 to 800 in 1999) may influence browsing on aspen, the composition and structure of aspen communities and (2) the long-term effects (up to 20-years-old post-clearcutting) that ungulate browsing on aspen communities can have on the habitat quality of wildlife communities and selective species of interest to wildlife and forest managers. Raymer concluded that browsing on bigtooth aspen in clearcuts 1–3 years old continued to be greater than quaking aspen clearcuts of the same age and that in aspen communities of both types that were older than 13 years old, browsing was causing a 50 percent reduction in merchantable timber volume, especially in stands 15–17 years old (Figures 2.4 and

2.5; 1999 photos). Ungulate browsing also reduced ruffed grouse habitat quality by reducing stem densities below thresholds that are suitable for grouse while habitat for the ovenbird was greater, especially in browsed quaking aspen communities. Raymer also documented aspen communities that sustained browsing of <50 percent on the current annual growth had different songbird species composition than aspen clearcuts browsed >50 percent. For example, 60 bird species were found in 36 aspen stands between 13–20 years old that were browsed <50 percent, whereas 46 bird species were found on stands browsed >50 percent. The ruffed grouse and white-throated sparrow (*Zonotrichia albicollis*) were both more abundant on stands with <50 percent browsing. Of the bird species found on these stands, 19 were uniquely found in aspen stands browsed <50 percent (e.g., golden-winged warbler [*Vermivora chrysoptera*], winter wren [*Troglodytes hiemalis*], black-throated green warbler [*Setophaga virens*]) whereas five species were only found in stands browsed >50 percent (e.g., eastern bluebird [*Sialia sialis*], field sparrow [*Spizella pusilla*], American woodcock).

Importance of Conserving the Range of Aspen Diversity for Biodiversity Conservation

Historically, forest management did not distinguish between aspen communities; aspen within regions was typically managed similarly. However, it is now evident that the structure and composition of aspen stands varies even within local scales due to differences in age, soil, hydrology, and disturbances. For instance, young aspen stands look different from older aspen stands, and aspen on sand looks different from aspen on loam. Evidence suggests that some aspen communities are not well represented in landscapes (Kessler et al. 1992) either because geographic areas supporting them are limited or because they are impacted by intense ungulate browsing or exploitation by humans. For example, if aspen harvesting strategies result in a 50-year age class not being represented in a specific habitat type, then potentially an ecological community may not be represented. Because of this knowledge, managers have realized the importance of conserving aspen diversity throughout landscapes as a means to conserve selected species and biodiversity at multiple spatial scales. Through harvesting activities, much of the aspen occurs in early successional stages, which provides important browse for herbivores and habitat for other species dependent on early successional vegetation. Conversely, managers may also let aspen succeed to later naturally occurring successional stages that could be composed of mixtures of oak, conifers, or hardwood species depending on soil types. With succession, aspen becomes a minor component of the composition of stands, but aspen trees can still provide essential habitat components for cavity-nesting birds (Dingledine and Haufler 1983) or flying squirrels (*Glaucomys* spp.; Holloway and Malcolm 2007) especially in landscapes where oaks or other hardwood species are preferred species for firewood removal and aspen is not (Dingledine and Haufler 1983). Maintaining representation of aspen in different age classes and habitat types subsequently maintains habitat for a diversity of wildlife communities.

In the Upper Peninsula of Michigan, distinct avian neotropical migrant communities associated with specific aspen age classes (20-years, 50–60 years, >70 years) and across a soil gradient represented by habitat types (Felix-Locher et al. 2010) were investigated. The study examined the relationship among avian communities and the structure and composition variables associated with aspen stands in different age classes and habitat types in a landscape without elk herbivory and relatively low deer densities. Essentially, avian species that may have been associated with the aspen community showed distinct associations with specific characteristics of aspen stands such as age or structure and composition of stands in different habitat types. For instance, suppose aspen diversity was represented two-dimensionally as a gradient between age and soil characteristics. Younger-aged aspen stands have higher deciduous stem densities and higher total stem densities than older stands; older stands have taller trees and larger-diameter trees (Figure 2.6). Drier, sandy aspen stands are characterized by having more conifers, higher percentage canopy cover, and lower stem densities than stands characterized by loamy and mesic-hydric soils. The higher moisture content and nutrient-rich soils of those stands support greater herbaceous and shrub cover than drier, sandier aspen stands.

Results of the study indicated that all aspen stands (bigtooth vs. quaking) are not ecologically equal, either between or within a species. That is, different aspen age classes and habitat types provide unique vegetation features important to different assemblages of avian communities (Felix-Locher et al. 2010). For example, species with strong associations with aspen communities in the 50-year age class on mesic loamy soils included the American redstart (*Setophaga ruticilla*), eastern peewee (*Contopus virens*), Northern flicker (*Colaptes auratus*), scarlet tanager (*Piranga olivacea*), and veery (*Catharus fuscescens*; Figure 2.6). These species prefer to nest in mature mesic hardwood forests with a dense understory (Brewer et al. 1991). Another community was found to occur in aspen older than 70 years supported by drier sandy soils. Those communities supported conifers (e.g., spruce and fir), which provide nesting and foraging habitat for associated birds including the black-and-white warbler (*Mniotilta varia*), brown creeper (*Certhia americana*), pine siskin (*Spinus pinus*), and yellow-rumped warbler (*Setophaga coronata*). If management is not maintaining some representation or threshold of a diversity of age classes across habitat types, wildlife communities associated with them will likewise not be represented. Further research should be initiated to determine what that threshold needs to be across a planning landscape to maintain viable wildlife communities across a range of age classes of various habitat types.

Investigation of ruffed-grouse use of aspen further supported the importance of understanding the importance of maintaining the ecological variability in aspen stands throughout a landscape for supporting conservation efforts. It is well known that aspen, especially young aspen, is important for providing ruffed-grouse drumming habitat in the Great Lakes and northern regions where aspen is prevalent (Gullion 1977; Kubisiak et al. 1980). However, aspen stands within different habitat types have different potentials to provide the vegetation structure necessary for drumming grouse during spring mating rituals. Specifically, young aspen in habitat types characterized by dry, sandy soils were less likely to be used by ruffed grouse for drumming than those characterized by mesic soils with a loam component or more hydric sites (Felix-Locher and Campa 2010).

In addition to aspen's ecological value to ungulates, selected game bird species, and diverse songbird communities, natural resource managers also attempt to provide even-aged forests for selected wildlife species or communities of conservation concern. Some of these management practices are implemented to influence primarily early successional stages and their associated wildlife communities while others are intended to influence later successional species or communities. For example, in Michigan, natural resource managers have implemented clearcutting treatments that have included leaving coarse woody debris on the forest floor and/or retaining canopy green trees (i.e., standing mature trees, often conifers) during clearcutting to emulate a diversity of ecological conditions after natural disturbances that may be beneficial to herptile species (e.g., red-backed salamanders [*Plethodon cinereus*]) and communities (Otto 2012) and potentially not be as visually unappealing to some stakeholders.

Evidence suggests that aspen is a key component to several communities and provides important habitat components for wildlife. It is clear that natural resource managers must consider the diversity of conditions on which aspen grows, how the trees respond to natural and anthropogenic disturbances, and maintain them as well as the diversity of age classes within those conditions. Without adequate representation of the historical variation of aspen across age classes and habitat types, the habitat availability for some wildlife species may be diminished. Furthermore, management efforts focused on managing for aspen may overlook factors that directly affect the presence of specific wildlife species or perhaps entire communities.

Managing Aspen for Diversity at Multiple Scales

Within the past decade or so, many state and federal agencies have focused attention on aspen conservation due to significant and sudden declines in many areas (e.g., sudden aspen decline in Colorado, see Worrall et al. 2008) and the increased understanding of the ecological and economic importance of aspen (Stelfox 1995; Piva 2003; Shepperd et al. 2006). Reported aspen age-class imbalances (Hammill and Visser 1984) in Michigan and other areas within the Great Lakes region have created a "boom and bust legacy problem for wildlife habitats and populations as well as the wood products industry" (Pedersen 2005: 21). With interest and objectives to conserve biodiversity, there is a need to create and maintain some representation or threshold of numerous, historically occurring aspen age classes over all habitat types capable of supporting aspen. A diversification of aspen age structure through management increases the resilience of landscapes to sudden declines (Gray and Mask 2009), provides habitat conditions for a diversity of wildlife communities, and sustains production. Intensive timber harvesting practices, such as those that harvest most of the aspen when it matures at 40–60 years, are not sustainable to maintain a constant supply of fiber or conserve the diversity of conditions important for wildlife (Locher et al. 2012). Rather, managers should defer harvest and allow aspen in some areas to reach naturally occurring old-growth or late successional conditions (i.e., large-diameter trees, cavities). Additionally, aspen should not be managed as a single forest type. The diversity of ecological conditions on which aspen is supported warrants consideration of habitat types in deciding when and where harvesting should take place to create a diversity of habitat conditions over broad landscapes.

Conclusion

Aspen communities that thrive following natural or anthropocentric disturbances can be perpetuated using strategic forest management practices that mimic the respective disturbances which would naturally occur spatially and temporally on the various habitat types in which aspen can be found. Historically, natural disturbances such as fires that were stand-replacing and created even-aged conditions maintained aspen. Today, in the eastern U.S. with the absence of using fire as a primary management practice or as a disturbance event, even-aged management practices (primarily clearcutting) are often used on a 40- to 60-year rotation, depending upon the habitat types in which the aspen community occurs, to emulate setting back succession under natural disturbances. Along with clearcutting, some state natural resource agencies elect to use prescriptions to retain conifers, snags, or seed trees within clearcuts to meet other wildlife or forest management objectives. Caution must be exercised, if too many conifers are retained in clearcuts;

these might provide hiding cover for ungulates to browse extensively and shade out shade-intolerant species such as aspen. Creating an abundance of early successional aspen to meet diverse forest and wildlife management objectives, however, can be problematic in the face of locally abundant ungulates using this resource to meet their habitat requirements. Successive years of ungulate browsing >50 percent of aspen current annual growth can be a significant enough disturbance to change the composition and structure of aspen communities, especially those on well-drained soil types, resulting in aspen mortality and declines in habitat quality for wildlife. In some instances, intensive browsing may interact with other local disturbances such as fire or extreme weather events, and aspen communities may not be able to recover if ecological conditions are beyond those to which they evolved. Such instances may be unpredictable, resulting in the inability to forecast outcomes in terms of aspen sustainability, resilience, or physiognomy.

Maintaining aspen communities for diverse timber and wildlife objectives in landscapes with ungulates occuring at densities that provide hunting and viewing opportunities will require an integrated management approach of (a) ungulate population management; (b) harvesting aspen in diverse rotation cycles that are appropriate for the disturbance cycles on the respective habitat types in which aspen communities occur; (c) understanding how future climate-change factors may influence aspen ecology on a variety of habitat types subjected to ungulate herbivory; and (d) the need to create a diversity of age classes of aspen communities throughout the planning landscape to disperse potential browsing pressure so it mimics the browsing pressures that aspen species most likely evolved with under lower ungulate densities than what we see in the twenty-first century.

References

Beyer, Jr., D. E. 1987. Population and habitat management of elk in Michigan. Ph.D. thesis, Michigan State University, East Lansing, Michigan, USA.

Brewer, R., G. A. McPeek, and R. J. Adams, Jr. 1991. *The Atlas of Breeding Birds of Michigan*. East Lansing, MI: Michigan State University Press.

Brinkman, K. A., and E. I. Roe. 1975. *Quaking Aspen: Silvics and Management in the Lake States*. Agricultural Handbook 486. Washington, D.C.: U.S. Department of Agriculture, Forest Service.

Burns, R. M., and B. H. Honkala (technical coordinators). 1990. *Silvics of North America: 2. Hardwoods*. Agriculture Handbook 654. Washington, D.C.: U.S. Department of Agriculture, Forest Service.

Burroughs, J. P., H. Campa III, S. R. Winterstein, B. A. Rudolph, and W. E. Moritz. 2006. Cause-specific mortality and survival of white-tailed deer fawns in southwestern lower Michigan. *Journal of Wildlife Management* 70: 743–751.

Campa, III, H. 1989. Effects of deer and elk browsing on aspen regeneration and nutritional qualities in Michigan. Ph.D. thesis, Michigan State University, East Lansing, Michigan, USA.

Campa, III, H., J. B. Haufler, and D. E. Beyer, Jr. 1992. Effects of simulated ungulate browsing on aspen characteristics and nutritional qualities. *Journal of Wildlife Management* 56: 158–164.

Campa, III, H., J. B. Haufler, and S. R. Winterstein. 1993. Effects of white-tailed deer and elk browsing on regenerating aspen: a ten year evaluation. In I. Thompson, editor, *Proceedings of the International Union of Game Biologists, XXI Congress: Forests and Wildlife: Towards the 21st Century*, 304–310. Halifax, Nova Scotia, Canada.

Campa, III, H., S. J. Riley, S. R. Winterstein, T. L. Hiller, S. A. Lischka, and J. P. Burroughs. 2011. Changing landscapes for white-tailed deer management in the 21st century: parcelization of landownership and evolving stakeholder values in Michigan. *Wildlife Society Bulletin* 35: 168–176.

Campa, III, H., D. K. Woodyard, and J. B. Haufler. 1986. Deer and elk use of forages treated with municipal sewage sludge. In D. W. Cole, C. L. Henry, and W. L. Nutter, editors, *The Forest Alternative for Treatment and Utilization of Municipal and Industrial Wastes*, 188–198. Seattle, WA: University of Washington Press.

Clausen, D. L., and R. F. Schroeder (compilers). 2004. Social acceptability of alternatives to clearcutting: discussion and literature review with emphasis on southeast Alaska. General Technical Report PNW-GTR-594, U.S. Department of Agriculture, Forest Service. Pacific Northwest Research Station, Portland, Oregon, USA.

Cleland, D. T., L. A. Leefers, and D. I. Dickmann. 2001. Ecology and management of aspen: a Lakes States perspective. In W. D. Shepperd, D. Binkley, D. L. Bartos, T. J. Stohlgren, L. G. Eskew, editors, *Proceedings of a Conference—Sustaining Aspen in Western Landscapes*, 81–99. General Technical Report RMRS-P-18. U.S. Department of Agriculture, Forest Service, Rocky Mountain Research Station, Fort Collins, Colorado, USA.

Davidson, R. W., T. E. Hinds, and F. G. Hawksworth. 1959. Decay of aspen in Colorado. Rocky Mountain Forest and Range Experiment Station, Station Paper 45. U.S. Department of Agriculture, Forest Service, Rocky Mountain Forest and Range Experiment Station, Fort Collins, Colorado, USA.

Daubenmire, R. 1966. Vegetation: Identification of typal communities. *Science* 151: 291–298.

Dickmann, D. I. 2006. Silviculture and biology of short-rotation woody crops in temperate regions: then and now. *Biomass Bioenergy* 30: 696–705.

Dickmann, D. I. and L. A. Leefers. 2004. *The Forests of Michigan*. Ann Arbor, MI: The University of Michigan Press.

Dingledine, J., and J. B. Haufler. 1983. The effect of firewood removal on breeding bird populations in a northern oak forest. In J. W. Davis, G. A. Goodwin, and R. A. Ockenfels, coordinators, *Snag Habitat Management: Proceedings of the Symposium*, 45–50. General. Technical Report RM-99. U.S. Department of Agriculture, Forest Service, Rocky Mountain Forest and Range Experiment Station, Fort Collins, Colorado, USA.

Felix, A. B. 2008. Modeling the cumulative effects of aspen management practices on timber and wildlife at multiple spatial and temporal scales. Ph.D. thesis, Michigan State University, East Lansing, Michigan, USA.

Felix, A. B., H. Campa III, K. F. Millenbah, S. R. Winterstein, and W. E. Moritz. 2004. Development of landscape-scale habitat-potential models for forest wildlife planning and management. *Wildlife Society Bulletin* 32: 795–806.

Felix-Locher, A., and H. Campa, III. 2010. Importance of habitat type classifications for predicting ruffed grouse use of areas for drumming. *Forest Ecology and Management* 259: 1464–1471.

Felix-Locher, A., H. Campa, III, and D. E. Beyer, Jr. 2010. Modeling avian community use of aspen following simulated harvest in Michigan. *Journal of Environmental Monitoring and Restoration* 6: 111–133.

Findell, V. E., R. E Pfeifer, A. G. Horn, C. H. Tubbs. 1960. Michigan's forest resources. Lake States Forest Experiment Station, Forest Service, U.S. Department of Agriculture, St. Paul, Minnesota, USA. https://archive.org/details/michigansforestr82find (accessed December 31, 2017).

Forest Inventory Data Online. 2015. DATIM: Design and Analysis Toolkit for Inventory and Monitoring. U.S. Department of Agriculture, Forest Service https://apps.fs.usda.gov/DATIM/Default.aspx? (Accessed February 10, 2019).

Frelich, L. E., and P. B. Reich. 1995. Spatial patterns and succession in a Minnesota southern boreal forest. *Ecological Monographs* 65: 325–346.

Frey, B. R., V. J. Lieffers, E. H. Hogg, and S. M. Landhäusser. 2004. Predicting landscape patterns of aspen dieback: Mechanisms and knowledge gaps. *Canadian Journal of Forest Research* 34: 1379–1390.

Graham, S. A., R. P. Harrison, Jr., and C. E. Westell, Jr. 1963. *Aspens, Phoenix Trees of the Great Lakes Region*. Ann Arbor, MI: University of Michigan Press.

Gray, S., and R. Mask. 2011. Sudden aspen declines in Colorado. Forest Health Protection. U.S. Department of Agriculture, Forest Service, Fort Collins, Colorado, USA. http://fs.usda.gov/Internet/FSE_DOCUMENTS/fsbdev3_038824.pdf (accessed February 10, 2019).

Gullion, G. W. 1977. Forest manipulation for ruffed grouse. *Transactions of the North American Wildlife and Natural Resources Conference* 42: 449–458.

Gullion, G. W., and F. J. Svoboda. 1972. The basic habitat resource for ruffed grouse. In *Aspen Symposium Proceedings*, 113–119. General Technical Report NC-1, U.S. Department of Agriculture, Forest Service, North Central Research Station, St. Paul, Minnesota, USA.

Hammill, J. H., and L. Visser. 1984. Status of aspen in northern Michigan as ruffed grouse habitat. Michigan Department of Natural Resources, Wildlife Division Report No 2976. Lansing, Michigan, USA.

Hart, J. H., J. B. Hart, and P. V. Nguyen. 1986. Aspen mortality following sludge application in Michigan. In D. W. Cole, C. L. Henry, and W. L. Nutter, editors, *The Forest Alternative for Treatment and Utilization of Municipal and Industrial Wastes*, 266–271. Seattle, WA: University of Washington Press.

Haines, D. A., and R. W. Sando. 1969. Climate conditions preceding historically great fires in the North Central Region. Research Paper NC-34. U.S. Department of Agriculture, Forest Service, North Central Forest Experiment Station, Minneapolis, Minnesota, USA.

Holloway, G. L., and J. R. Malcolm. 2007. Nest-tree use by northern and southern flying squirrels in central Ontario. *Journal of Mammalogy* 88: 226–233.

Holste, E., and B. Garmon. 2013. Managing Michigan's state-owned forests: harvest levels, market trends and revenue realities. Michigan Environmental Council. Lansing, Michigan, USA. https://www.environmentalcouncil. org/mecReports/ManagingMichigansState-ownedForests.pdf (accessed December 31, 2017).

Hummel, S. L., H. Campa III, S. R. Winterstein, and E. M. Dunton. 2018. Understanding how a keystone herbivore, white-tailed deer impacts wetland vegetation types in southern Michigan. *The American Midland Naturalist* 179: 51–67.

Iverson, L. R., and A. M. Prasad. 1998. Predicting abundance of 80 tree species following climate change in the Eastern United States. *Ecological Monographs* 68: 465–485.

Johnston, R. S., and R. D. Doty. 1972. Description and hydrologic analysis of two small watersheds in Utah's Wasatch Mountains. USDA Forest Service Research Paper INT-127, Intermountain Forest and Range Experiment Station, Ogden, Utah, USA.

Jones, J. R., and N. V. DeByle. 1985. Fire. In N. V. DeByle and R. P. Winokur, editors, *Aspen: Ecology and Management in the Western United States*, 77–81. Forest Service, General Technical Report RM-119. US Department of Agriculture, Forest Service, Rocky Mountain Forest and Range Experiment Station, Fort Collins, Colorado, USA.

Kessler, W. B., H. Salwasser, C. W. Cartwright, Jr., and J. A. Caplan. 1992. New perspectives for sustainable natural resources management. *Ecological Applications* 2: 221–225.

Kohn, B. E., and J. J. Mooty. 1971. Summer habitat of white-tailed deer in north-central Minnesota. *Journal of Wildlife Management* 35: 476–487.

Kotar, J., and T. L. Burger. 2000. *Field Guide to Forest Habitat Type Classification for North Central Minnesota*. Madison, WI: Terra Silva Consultants.

Kubisiak, J. F., J. C. Moulton, and K. R. McCaffery. 1980. Ruffed grouse (*Bonasa umbellus*) density and habitat relationships in Wisconsin, USA. *Wisconsin Department of Natural Resources Technical Bulletin* 189: 1–29.

Locher, A., H. Campa, III, L. A. Leefers, and D. E. Beyer, Jr. 2012. Understanding cumulative effects of aspen harvest on wildlife habitat and timber resources in northern Michigan. *Northern Journal of Applied Forestry* 29: 113–127.

Lovely, K. R., W. J. McShea, N. W. Lafron, and D. E. Carr. 2013. Land parcelization and deer population densities in a rural county in Virginia. *Wildlife Society Bulletin* 37: 360–367.

McCabe, T. R., and R. E. McCabe. 1997. Recounting whitetails past. In W. J. McShea, H. B. Underwood, and J. H. Rappole, editors, *Deer Ecology and Population Management*, 11–26. Washington, D.C.: Smithsonian Institution Press.

McShea, W. J. 2012. The ecology and management of white-tailed deer in a changing world. Annals of New York Academy of Sciences. *The Year in Ecology and Conservation Biology* 1249: 45–56.

McShea, W. J., H. B. Underwood, and J. H. Rappole (editors). 1997. *The Science of Overabundance. Deer Ecology and Population Management*. Washington, D.C.: Smithsonian Institution Press.

Michigan Department of Natural Resources. 1975. *State of Michigan Environmental Impact Statement for Potential Hydrocarbon Development on the Pigeon River*

Country State Forest, Montmorency, Otsego, and Presque Isle Counties. Lansing, MI: Michigan Department of Natural Resources.

Michigan Department of Natural Resources. 2018. Elk updates: results of 2017 elk hunting season, winter elk survey. https://www.michigan.gov/som/0,4669,7-192-47796-459787-,00.html (accessed July 1, 2018).

Moran, R. 1973. *The Rocky Mountain Elk in Michigan.* Research and Development Report 267. Lansing, MI: Michigan Department of Natural Resources.

Mueggler, W. F. 1985. Vegetation associations. In N. V. DeByle, and R. P. Winokur, editors, *Aspen: Ecology and Management in the Western United States*, 45–55. General Technical Report RM-19. U.S. Department of Agriculture, Forest Service, Fort Collins, Colorado, USA.

Murie, O. J. 1951. *The Elk of North America.* Washington, D.C.: The Wildlife Management Institute, Stackpole Books.

Oswalt, S. N., W. B. Smith, P. D. Miles, and S. A. Pugh. 2014. Forest resources of the United States, 2012: a technical document supporting the Forest Service update of the 2010 RPA assessment. USDA Forest Service General Technical Report WO-91, Washington, D.C.., USA.

Otto, C. R. V. 2012. The impact of timber harvest on wildlife distribution patterns and population vital rates: does structural retention ameliorate the negative effects of clearcutting. Ph.D. thesis. Michigan State University. East Lansing, Michigan, USA.

Palik, B. J. and K. S. Pregitzer. 1992. A comparison of presettlement and present-day forests on two bigtooth aspen-dominated landscapes in Northern Lower Michigan. *The American Midland Naturalist* 127: 327–338.

Parsons, S., K. J. Lewis, and J. M. Psyllakis. 2003. Relationships between roosting habitat of bats and decay of aspen in the sub-boreal forests of British Columbia. *Forest Ecology and Management* 177: 559–570.

Pedersen, L. 2005. Michigan state forest timber harvest trends. A review of recent harvest levels and factors influencing future levels. Michigan Department of Natural Resources, Forest Management Division, Lansing, Michigan, USA.

Perala, D. A. 1990. *Quaking Aspen.* In R. M. Burns and B. H. Honkala, technical coordinators, *Silvics of North America: 2. Hardwoods 2.* Agriculture Handbook 654. Washington, D.C.: U.S. Department of Agriculture, Forest Service.

Piva, R. J. 2003. Pulpwood production in the north-central region, 2000. Resource Bulletin NC-221. US Department of Agriculture, Forest Service, North Central Research Station. St. Paul, Minnesota, USA.

Piva, R. J. 2007. Pulpwood production in the North-Central Region. 2005. Research Bulletin NRS-21. U.S. Department of Agriculture, Forest Service, Northern Research Station, Minneapolis, Minnesota, USA.

Pojar, R. A. 1995. Breeding bird communities in aspen forests of the sub-boreal spruce (dk subzone) in the Prince Rupert Forest Region. *Forest Management Handbook*, Chapter 33. Province of British Columbia, Ministry of Forest Research Program, Victoria, B.C., Canada.

Raymer, D. 2000. Effects of elk and white-tailed deer browsing on aspen communities and wildlife habitat quality in northern lower Michigan: an 18-year evaluation. Ph.D. thesis, Michigan State University, East Lansing, Michigan, USA.

Rogers, P. C., S. M. Landhausser, B. D. Pinno, and R. J. Ryel. 2014. A functional framework for improved management of western North American aspen (*Populus tremuloides* Michx.). *Forest Science* 60: 345–355.

Rooney, T. P., and D. M. Waller. 2003. Direct and indirect effects of white-tailed deer in forest ecosystems. *Forest Ecology and Management* 181: 165–176.

Schier, G. A., and A. D. Smith. 1979. Sucker regeneration in a Utah aspen clone after clearcutting, partial cutting, scarification, and girdling. Research Note INT-253. U.S. Department of Agriculture, Forest Service, Intermountain Forest and Range Experiment Station, Ogden, Utah, USA.

Schmitt, S. M., D. J. O'Brien, C. S. Bruning-Fann, and S. D. Fitzgerald. 2002. Bovine tuberculosis in Michigan wildlife and livestock. *Annals of the New York Academy of Sciences* 969: 262–268.

Seton, E. T. 1909. *Life History of Northern Mammals*, Vol. I. New York: Charles Scribner's Sons.

Seton, E. T. 1929. *Lives of Game Animals*, Vol. III. Part 1. Garden City, NY: Doubleday, Doran and Company.

Shepperd, W. D. 1981. Stand characteristics of Rocky Mountain aspen. In N. V. DeByle, editor, *Situation Management of Two Intermountain Species: Aspen and Coyotes. Symposium Proceedings. Volume I, Aspen*. Logan, UT: Utah State University.

Shepperd, W. D., P. C. Rogers, D. Burton, D. L. Bartos. 2006. Ecology, biodiversity, management, and restoration of aspen in the Sierra Nevada. General Technical Report RMRS-GTR-178. U.S. Department of Agriculture, Forest Service, Rocky Mountain Research Station, Fort Collins, Colorado, USA.

Spencer, J. S., Jr. and H. W. Thorne. 1972. Wisconsin's 1968 timber resource. Resource Bulletin NC-15. U.S. Department of Agriculture, Forest Service, St. Paul, Minnesota, USA.

Stearns, F. 1995. History of Wisconsin's northern forests and the pine barrens. In E. A. Borgerding, G. A. Bartelt, and W. M. McCowen, editors, *The Future of Pine Barrens in Northwestern Wisconsin: A Workshop Summary: Proceedings of the Workshop*, 4–6; 21–23 September 1993. Madison, WI: Wisconsin Department of Natural Resources.

Stelfox, J. B. (editor). 1995. *Relationships between Stand Age, Stand Structure, and Biodiversity in Aspen Mixed Wood Forest in Alberta*. Edmonton, Alberta, Canada: Alberta Environmental Center and Canadian Forest Service.

Urbanek, R. E., C. K. Nielsen, G. A. Glowacki, and T. S. Pruess. 2012. Effects of white-tailed deer (*Odocoileus virginianus* Zimm.) herbivory in restored forests and savanna plant communities. *The American Midland Naturalist* 167: 240–255.

Westell, Jr., C. E. 1954. Available browse following aspen logging in lower Michigan. *Journal of Wildlife Management* 18: 266–271.

Worrall, J. J., L. Egeland, T. Eager, R. A. Mask, E. W. Johnson, P. A. Kemp, and W. D. Shepperd. 2008. Rapid mortality of *Populus tremuloides* in southwestern Colorado, USA. *Forest Ecology and Management* 225: 686–969.

Zasada, J. C., and H. M. Phipps. 1990. Balsam poplar. In R. M. Burns and B. H. Honkala, technical coordinators, *Silvics of North America: Hardwoods 2*, 518–529. Agriculture Handbook 654. Washington, D.C.: U.S. Department of Agriculture, Forest Service.

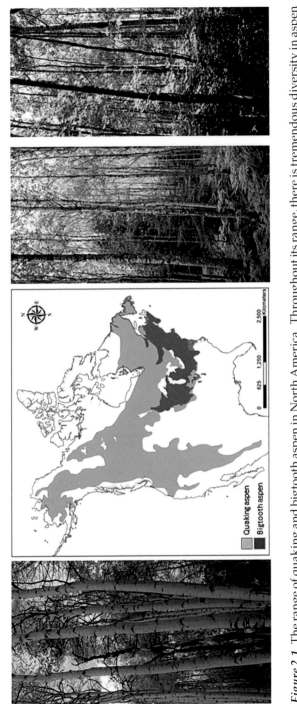

Figure 2.1 The range of quaking and bigtooth aspen in North America. Throughout its range, there is tremendous diversity in aspen community composition and structure depending on climate, soils, and geology. Photos illustrate differences in aspen community structure in Colorado (left) and on xeric sandy and mesic loamy soils in Michigan (right). (Photo credits: (left) Pixabay-853565; (right) A. Felix.)

BTA #72; 4yr old; Exclosure 1986   BTA #72; 4yr old; Open to Browsing 1986

BTA #72; 17yr old; Exclosure 1999   BTA 17yr old; Open to Browsing 1999

Figure 2.4 Stand structure of a 4-year-old bigtooth aspen (BTA; numbers correspond to stand location) community within a 20 × 20 m exclosure and a paired area open to white-tailed deer and elk browsing in northern lower Michigan in 1986. All exclosures were built on clearcuts immediately following clearcutting in 1982. The stand structure of the same aspen community within the exclosure and paired area 13 years later. (Photo credits: (top left and right) H. Campa III, 1986; (bottom left and right) D. Raymer, 1999.)

QA #14; 6yr old; Exclosure 1986   QA#14; 6yr old; Open to Browsing 1986

QA #14; 19yr old; Exclosure 1999   QA #14; 19yr old; Open to Browsing 1999

Figure 2.5 Stand structure of a 6-year-old quaking aspen (QA; numbers correspond to stand location) community within a 20 × 20 m exclosure and a paired area open to white-tailed deer and elk browsing in northern lower Michigan in 1986. The stand structure of the same aspen community within the exclosure and paired area 13 years later. (Photo credits: (top left and right) H. Campa III, 1986; (bottom left and right) D. Raymer, 1999.)

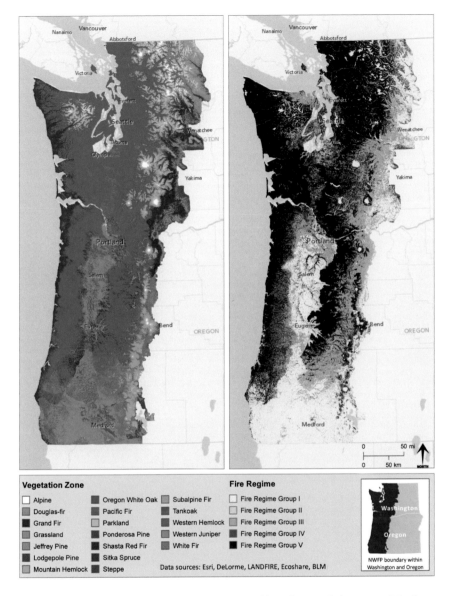

Vegetation Zone

Alpine
Douglas-fir
Grand Fir
Grassland
Jeffrey Pine
Lodgepole Pine
Mountain Hemlock
Oregon White Oak
Pacific Fir
Parkland
Ponderosa Pine
Shasta Red Fir
Sitka Spruce
Steppe
Subalpine Fir
Tankoak
Western Hemlock
Western Juniper
White Fir

Fire Regime

Fire Regime Group I
Fire Regime Group II
Fire Regime Group III
Fire Regime Group IV
Fire Regime Group V

Data sources: Esri, DeLorme, LANDFIRE, Ecoshare, BLM

NWFP boundary within
Washington and Oregon

Figure 3.1 Modeled vegetation zones (http://ecoshare.info/category/gis-data-vegzones/) corresponding to fire regime groups (https://www.landfire.gov/) within the Northwest Forest Plan (NWFP) boundary in Oregon and Washington based on the Interagency Clearinghouse of Ecological Information dataset for 19 modeled vegetation zones (http://ecoshare.info/category/gis-data-vegzones/). Fire regime groups include: I (0 to 35-year frequency, low to mixed severity); II (0 to 35-year frequency, replacement severity); III (35 to 200-year frequency, low to mixed severity); IV (35 to 200-year frequency, replacement severity); and V (200+ year frequency, any severity). (Figure prepared by Jessica Leonard, Geos Institute.)

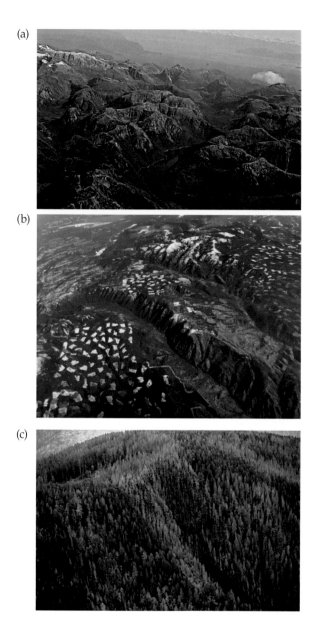

Figure 3.2 Contrasting landscapes with different legacy functions as viewed from an airplane window: (a) large landscape with intact and well-connected low- to mid-elevation forests in different successional stages integrated with high-elevation alpine areas in the Pacific Northwest. (Photo credit: D. DellaSala.) (b) highly fragmented pattern of dispersed clearcuts and homogeneous plantations in East Kooney Mountains, BC. (Photo credit: D. DellaSala.) (c) fire mosaic (pyrodiversity) in southwest Oregon (Biscuit fire 2002) showing patch heterogeneity. (Photo credit: K. Crocker.)

Figure 3.3 Large snags originating from a disturbance at the site level create complex structures that link successional stages across time and space. (Photo credit: D. DellaSala, Biscuit fire 2002, southwest Oregon, taken 10 years post-fire.) (See Donato et al. 2012.)

Figure 3.4 A large "nurse log" left by a fallen tree provides a substrate for plants that eventually in-fill the canopy gap, while the log acts as a microclimate site for salamanders and numerous invertebrates, which are especially important in dry summer months. (Photo credit: D. DellaSala, coast redwoods, *Sequoia sempervirens*, southern Oregon.)

Figure 3.5 A massive coast redwood in Oregon that survived fires as evident by the large fire-created cavity in the tree center. Legacies link human generations to the natural world through a sense of appreciation and intrinsic value for wild places. (Photo credit: D. DellaSala.)

(a)

Figure 3.6 (a) Large fires in dry forests include severely burned patches that generate complex-early seral forests characterized by abundant structures (snags, logs). (Photo credit: D. DellaSala, East Antelope fire, southwest Oregon, 13 years post-fire.) (b) Post-fire logging removes most (if not all) of the biological legacies on site. (Photo credit: D. DellaSala, Biscuit burn area, southwest Oregon.)

continued

(b)

Figure 3.6 (a) Large fires in dry forests include severely burned patches that generate complex-early seral forests characterized by abundant structures (snags, logs). (Photo credit: D. DellaSala, East Antelope fire, southwest Oregon, 13 years post-fire.) (b) Post-fire logging removes most (if not all) of the biological legacies on site. (Photo credit: D. DellaSala, Biscuit burn area, southwest Oregon.)

Figure 3.7 Google Earth image of intensively logged and fragmented landscape just before a large fire event (red fire perimeter) in southwest Oregon. The Douglas fire of 2013 burned intensely when it encountered numerous tree plantations (light green areas) during extreme fire weather (see Zald and Dunn 2018).

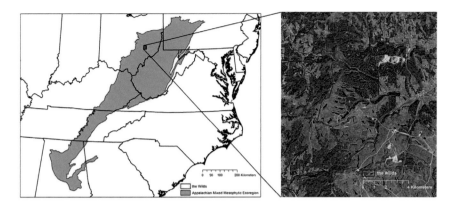

Figure 6.1 Location of The Wilds within the Appalachian mixed mesophytic forest ecoregion in Ohio. Appalachian mixed mesophytic forest ecoregion map adapted from Olson et al. (2001).

Figure 6.3 A large-scale ecological research project focused on understanding how native pollinators interact with a landscape that has been disturbed from mining and now is recovering through reclamation and restoration efforts. This work illustrates the stability and robustness of plant–pollinator interactions that can be achieved by introducing diverse prairie and meadow plants, even in the face of isolation from remnant habitat. (Photo credit: Joe Clark.)

chapter three

Fire-mediated Biological Legacies in Dry Forested Ecosystems of the Pacific Northwest, USA

Dominick A. DellaSala

What are Biological Legacies?

Biological legacies can be defined as the type, quantities, and patterns of biotic structures and organisms present before a disturbance that transfer their functions over to the post-disturbance environment (Franklin et al. 2002; Dale et al. 2005). Large (landscape-scale) fires of mixed-severity effects within dry-fire-adapted Ponderosa pine (*Pinus ponderosa*) and mixed-conifer forests of the Pacific Northwest USA, the subject of this chapter, generate pulses of biological legacies that can support ecosystems for decades to centuries (Franklin et al. 2000; Lindenmayer et al. 2008; Swanson et al. 2011; Donato et al. 2012; DellaSala and Hanson 2015). Whereas forest managers have focused mainly on legacies at the stand level, legacies also exist at landscape scales and as plant and wildlife populations persisting before and after a disturbance. Therefore, scale and context matter in managing legacies for biodiversity benefits.

In this chapter, my main objectives are to summarize the importance of forest legacies at multiple spatiotemporal scales, describe the ecosystem benefits of large fires that regularly accrue legacies in space and time in these forests, and distinguish natural disturbance events (fire) from chronic ones caused by intensive land management (e.g., roads and short-rotation logging). I also provide ways to maintain legacies in areas intensively managed for timber production and where large-landscape conservation is emphasized. I focus on fires of mixed-severity effects on dry forests, as a principal disturbance agent for legacy replenishment; other natural disturbances are not covered here. Mixed-severity fires generate a characteristic burn mosaic pattern in low- to mid-elevation dry forests in this region. The fire mosaic includes variably sized patches of lightly burned, moderately burned, and severely burned patches (DellaSala and Hanson 2015). This type of fire-mediated mosaic has been described as "pyrodiversity" and is associated with high levels of biodiversity in these forests (i.e., pyrodiversity begets biodiversity in this case, Parr and Brockett 1999; DellaSala and Hanson 2015). I focus on dry forests within the Northwest Forest Plan (NWFP) area of the Pacific Northwest, where extensive research on biological legacies has occurred (Franklin et al. 2000, 2002; Dale et al. 2005; Fontaine et al. 2009; Swanson et al. 2011; Donato et al. 2012; Odion et al. 2014), and where there is conservation interest in maintaining ecosystem benefits of large fires (DellaSala and Hanson 2015; also see DellaSala et al. 2017 for the Sierra region). Notably, the NWFP shifted federal land management from focusing primarily on timber production to focusing on ecosystem management and biodiversity conservation, across 10 million ha from northern California to Washington. The plan is considered a global model for large-landscape conservation, including maintenance of natural disturbance processes (DellaSala et al. 2015a).

Vegetation Types and Fire Regimes of the Pacific Northwest

Fire regime groups for this region ranged from frequently occurring (i.e., every 0–35 years) fires with low-severity effects on vegetation (i.e., most trees survive) to infrequent (>200 years) fires with high-severity (stand replacement) effects (i.e., most trees fire killed; Figure 3.1). Multiple lines of evidence indicate the preponderance of mixed-severity fire effects in dry forests of this region and the biodiversity importance of high-severity burn patches (>70 percent overstory mortality) within fire complexes (Hessburg et al. 2006; Perry et al. 2011; Baker 2012, 2015; Donato et al. 2012; Williams and Baker 2012; Odion et al. 2014). Importantly, the proportion and extent of high-severity fire has remained relatively constant from 1984–2011 in this region, replenishing legacies at landscape scales (Odion et al. 2014; Law and Waring 2015).

Mixed-severity fires generate pulses of biological legacies that result in structurally complex, early-seral forests (Swanson et al. 2011). Although fires of generally low intensity may generate patches of high tree mortality (fine-scale heterogeneity) in small-scale flare-ups, they do not produce the landscape heterogeneity characteristic of mixed-severity fires. It is this precise variation in patch patterns that sets fires of mixed-severity effects apart from fires of low-severity effects in terms of legacy production.

Legacies

Landscape

When viewed at the landscape scale, a collection of intact forest blocks in different post-disturbance stages can act as a legacy landscape. The

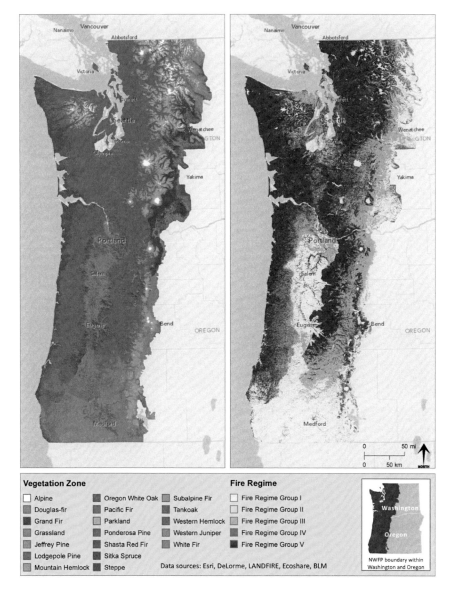

Figure 3.1 **(See color insert.)** Modeled vegetation zones (http://ecoshare.info/category/gis-data-vegzones/) corresponding to fire regime groups (https://www.landfire.gov/) within the Northwest Forest Plan (NWFP) boundary in Oregon and Washington based on the Interagency Clearinghouse of Ecological Information dataset for 19 modeled vegetation zones (http://ecoshare.info/category/gis-data-vegzones/). Fire regime groups include: I (0 to 35-year frequency, low to mixed severity); II (0 to 35-year frequency, replacement severity); III (35 to 200-year frequency, low to mixed severity); IV (35 to 200-year frequency, replacement severity); and V (200+ year frequency, any severity). (Figure prepared by Jessica Leonard, Geos Institute.)

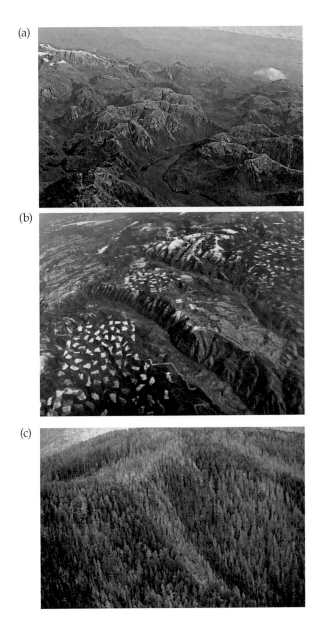

Figure 3.2 **(See color insert.)** Contrasting landscapes with different legacy functions as viewed from an airplane window: (a) large landscape with intact and well-connected low- to mid-elevation forests in different successional stages integrated with high-elevation alpine areas in the Pacific Northwest. (Photo credit: D. DellaSala.) (b) highly fragmented pattern of dispersed clearcuts and homogeneous plantations in East Kooney Mountains, BC. (Photo credit: D. DellaSala.) (c) fire mosaic (pyrodiversity) in southwest Oregon (Biscuit fire 2002) showing patch heterogeneity. (Photo credit: K. Crocker.)

heterogeneity present within intact areas is most notably distinguishable from the more-homogeneous human-dominated landscape (Figure 3.2a vs. 3.2b). Most of the Pacific Northwest is in a highly fragmented and altered condition, consisting of few natural legacy blocks interspersed within a highly industrially managed matrix (Heilman et al. 2002; Strittholt and DellaSala 2001). Highly fragmented landscapes have highly altered legacy functions, and the fragmented patches are mostly edge environments (Chen et al. 1992) that degrade habitat value for many plants and wildlife species, although individual patches, such as roadless areas embedded within the human-dominated matrix, maintain important legacy functions that may serve as "blueprints" for restoring larger blocks (Strittholt and DellaSala 2001). Legacy landscapes also include a mix of undisturbed and naturally disturbed sites that provide habitat for a suite of species distributed across the disturbance gradient (Figure 3.2c; Mackey et al. 2002; Thompson et al. 2007; Lindenmayer et al. 2008). Interestingly, what may function as an undisturbed patch for disturbance-avoiding species, may not for disturbance-adapted ones until the patch is perturbed, rendering it temporarily unsuitable for disturbance avoiders. Therefore, heterogeneity is important for maintaining biodiversity at the landscape scale.

Stand

Legacies produced by small-scale disturbances such as tree fall gaps or a flare-up within a predominately low-intensity burn provide fine-scale heterogeneity and production of localized legacies. Examples include large snags, downed logs, below-ground seed banks, and flowering plants that were present pre-disturbance and that persist in the newly created disturbed patch (Figure 3.3). Stand legacies perform myriad ecosystem functions as keystone structures (Franklin et al. 2000, 2002; Dale et al. 2005, Box 3.1).

BOX 3.1   STAND-LEVEL AND POPULATION LEGACIES

Many species of insects, birds, mammals, and herpetofauna rely on dead wood (snags and logs) for nesting, roosting, foraging, and hiding. In fact, some 67 species of vertebrates in this region are strongly dependent (i.e., use them >50 percent of the time) on snags (Bunnell 2013). The root wads of snags anchor soils, preventing erosion. Large snags shade conifer seedlings from intense sunlight, aiding in reforestation. Downed logs absorb and slowly release runoff, which is especially important to moisture-dependent

bryophytes, salamanders, terrestrial mollusks, and many invertebrates in dry summer months. Logs decompose slowly and
recycle soil nutrients while acting as "nurseries" for new plant
growth (Figure 3.4). Wildlife populations present before the disturbance also may act as source populations, post-disturbance
(Lindenmayer et al. 2005). Animal carcasses connect the deceased
with the living (namely, scavengers) as a food legacy. Interestingly,
spawned-out salmon (*Oncorhynchus* spp.) carcasses replenish soil
nutrients for plants in riparian areas (Reimchen 2000).

From a social perspective, forest legacies connect human generations to
the natural world in a way that is not easily quantifiable but important in
providing solitude, wilderness, and a sense of the great outdoors (Figure
3.5). A hike in old-growth forest can provide (1) a cultural connection to
previous forest conditions, and (2) perspective about how ecological base-
lines have shifted over time as forests continue to change under natural and
anthropogenic disturbance regimes.

Figure 3.3 **(See color insert.)** Large snags originating from a disturbance at the
site level create complex structures that link successional stages across time and
space. (Photo credit: D. DellaSala, Biscuit fire 2002, southwest Oregon, taken 10
years post-fire.) (See Donato et al. 2012.)

Figure 3.4 **(See color insert.)** A large "nurse log" left by a fallen tree provides a substrate for plants that eventually in-fill the canopy gap, while the log acts as a microclimate site for salamanders and numerous invertebrates, which are especially important in dry summer months. (Photo credit: D. DellaSala, coast redwoods, *Sequoia sempervirens*, southern Oregon.)

Figure 3.5 **(See color insert.)** A massive coast redwood in Oregon that survived fires as evident by the large fire-created cavity in the tree center. Legacies link human generations to the natural world through a sense of appreciation and intrinsic value for wild places. (Photo credit: D. DellaSala.)

Legacies Link Successional Pathways

> When we try to pick out anything by itself, we find
> it hitched to everything else in the universe.
>
> **(John Muir)**

The quantity and quality of legacies at the stand and landscape levels depend on what was present before the disturbance. In this fashion, the post-disturbance environment is "hitched" to the quality of its predecessor stage and vice versa (Donato et al. 2012). For instance, an intense fire in a mature forest triggers the onset of the complex pioneering stage of forests characterized by abundant legacies that then set the stage for development of older forest conditions over time (Swanson et al. 2011; Figure 3.3 and 3.6a). This early stage can persist for up to two decades (depending on site productivity) until forest canopy closure; however, it rarely persists in areas with post-fire logging (Figure 3.6b). The early stage is often disrupted by intense forest management as discussed below.

(a)

Figure 3.6 **(See color insert.)** (a) Large fires in dry forests include severely burned patches that generate complex-early seral forests characterized by abundant structures (snags, logs). (Photo credit: D. DellaSala, East Antelope fire, southwest Oregon, 13 years post-fire.) (b) Post-fire logging removes most (if not all) of the biological legacies on site. (Photo credit: D. DellaSala, Biscuit burn area, southwest Oregon.)

continued

(b)

Figure 3.6 **(See color insert.)** (a) Large fires in dry forests include severely burned patches that generate complex-early seral forests characterized by abundant structures (snags, logs). (Photo credit: D. DellaSala, East Antelope fire, southwest Oregon, 13 years post-fire.) (b) Post-fire logging removes most (if not all) of the biological legacies on site. (Photo credit: D. DellaSala, Biscuit burn area, southwest Oregon.)

Pulse vs. Chronic Disturbances

Disturbance can be characterized as having pulse (rapid) or chronic (long-term) effects on ecosystems, depending on intensity and frequency (Frelich 2008). For instance, legacies are continuously replenished by reoccurring pulse disturbances, the timing of which is governed by localized natural disturbance dynamics. If the fire return interval is over a century, the later stages of forest succession (old growth) may ensue over time with legacies continuously replenished at the stand level by small-scale events (e.g., gap-phase dynamics). Additionally, large fires are the main event for perpetuating abundant quantities of legacies at landscape scales. Many ecosystems and species evolved with these types of pulse disturbances (at stand and landscape), and have unique adaptations to cope with and even thrive in the post-disturbance environment (Box 3.2).

BOX 3.2 DISTURBANCE ADAPTATIONS

Many native species have unique adaptations that allow them to prosper, resist, and/or be resilient to natural disturbances. The thick bark of large, older, fire-resistant trees, such as Ponderosa pine, insulate the tree from low- to moderate-fire intensities. The giant sequoia (*Sequoiadendron giganteum*) is considered the most fire-resistant conifer on Earth, capable of withstanding even crown fires. Other species not only are resilient to disturbed areas but prosper in them as well. Examples include serotinous cones of certain lodgepole pine (*P. contorta*) populations that release their seeds after intense fire heat. Below-ground seed propagules of some shrub species (e.g., *Ceanothus* spp.) also require intense heat to crack open the seed coat. Stump- sprouting in madrone (*Arbutus menziesii*) and coast redwood are natural adaptations to disturbances. Many animal species are well adapted to colonize disturbed areas. The Black-backed Woodpecker (*Picoides arcticus*), aptly named for its cryptically colored back against charred tree boles, reaches its highest population levels in severely burned forests with abundant legacies that support beetle larvae, the woodpecker's predominant food source (Hutto 1995). Some wood-boring beetles have specialized infrared receptors (pit organs, e.g., *Melanophila* spp.) for detecting heat or smoke as far away as 60 km (Schmitz and Bousack 2012). These "fire-chasers" quickly hone in on fire-killed trees and are essential to nutrient cycling, post-disturbance.

In contrast, chronic disturbances from industrial-scale logging and road building often differ from pulse disturbances in cumulative effects to ecosystems. Such disturbances often overlap and accumulate in space and time (e.g., post-fire logging, roads, grazing, and mining, of which two or more may occur on the same site at the same time), which can flip ecosystem dynamics and species composition to highly altered (novel) states (Paine et al. 1998; Lindenmayer et al. 2011, 2017; Box 3.3; Figure 3.7). In such cases, a homogenized landscape that lacks legacies and biodiversity can result.

BOX 3.3 CHRONIC DISTURBANCES

Extensive logging and road building following fire is common in the Pacific Northwest (Figure 3.7; DellaSala et al. 2015b). Acting synergistically, several linked anthropogenic disturbances can often exceed the evolutionary capacity of species and ecosystems to respond to cumulative disturbances (Paine et al. 1998). An example comes from the highly fragmented landscapes of the Klamath-Siskiyou ecoregion of southwest Oregon and northern California, where industrial-scale logging has contributed to uncharacteristically severe fires because of the accumulation of logging slash and small, densely stocked and flammable trees that have replaced native fire-resistant tree species (Odion et al. 2004; Zald and Dunn 2018) (Figures 3.2b, 3.7). Most of the dead and live legacy trees are removed and this is followed by planting trees that often burn intensely in the next fire (Thompson et al. 2007). The logged area sometimes includes livestock grazing, colonization by invasive plants, and off-highway vehicle use (soil compaction) that can exceed disturbance thresholds of many native species (see Lindenmayer et al. 2011, 2017). Such landscape-scale conversions set in motion a management feedback loop, whereby a natural disturbance regime is transformed to frequently occurring and cumulative disturbances with concomitant losses in biodiversity and forest legacies (Lindenmayer et al. 2011, 2017; DellaSala et al. 2015b).

Figure 3.7 (**See color insert.**) Google Earth image of intensively logged and fragmented landscape just before a large fire event (red fire perimeter) in southwest Oregon. The Douglas fire of 2013 burned intensely when it encountered numerous tree plantations (light green areas) during extreme fire weather (see Zald and Dunn 2018).

Large Fires and Biological Legacies

Landscape-scale fires occur infrequently (from decades to centuries) in forest ecosystems of this region, but have disproportionate effects on ecosystems compared with small or low-intensity fires. Large fires of varied patch sizes and severities include severely burned patches with hyperabundant legacies (Figure 3.2c, Figures 3.3 and 3.6). Such fires result in high levels of alpha (within-stand) and beta diversity (species turn over across gradients) because of complex structures present at the stand level and varied patch severities at the landscape scale (Hanson et al. 2015). In general, prescribed fires and low-intensity burns lack this heterogeneity (Swanson et al. 2011; DellaSala et al. 2017).

Notably, under "extreme" weather conditions (e.g., drought, high winds, high temperatures), fires are likely to reach landscape proportions and therefore result in a pulse of legacies. Drought monitoring, in particular, is useful in predicting the likelihood of large fires and therefore where legacy pulses may occur. Managers of national parks, for instance, may want to know when and where there is potential for a major-fire event headed their way so that they can manage the fire for ecosystem benefits. US drought-monitor maps (https://droughtmonitor.unl.edu) have commonly been used to alert fire suppression forces, but not to predict where ecosystem benefits may accrue from large fires.

Landscape Conservation and Disturbance Dynamics

Landscape-scale conservation requires knowledge of disturbance dynamics in conservation biology approaches (Noss and Cooperrider 1994; DellaSala et al. 2017). In mixed-severity fire regimes, this means developing a conservation strategy that:

1. Includes a protected-areas network representative of patch types (e.g., fire severities) well-distributed in redundant patterns that collectively are resilient to natural disturbances (i.e., the disturbance does not overwhelm the functionality of the reserve network given redundant types, DellaSala et al. 2015a, 2017);
2. Maintains viable populations of disturbance-adapted and disturbance-avoiders, well distributed across a planning area; and
3. Provides landscape connectivity of pre- and post-disturbed areas to allow dispersal of plants and wildlife, particularly in a changing climate.

The reserve network of the NWFP incorporated resilience concepts into the design of the network. For instance, late-successional reserves (LSRs) are well distributed and in redundant patterns, so the loss of any particular LSR(s) should not overwhelm the integrity of the overall reserve network (DellaSala et al. 2015a). However, very few reserves are big enough to fully encompass large fires and therefore managing fire for ecosystem benefits across landscapes needs to include both protected and other land-use designations. For example, although 71,351 ha (39 percent) of the 184,354 ha 2002 Biscuit fire perimeter in southwest Oregon occurred within the Kalmiopsis Wilderness, extensive fire suppression was employed throughout the burn area, bulldozed fire lines created travel corridors for weedy species, post-fire logging damaged LSRs and encroached on wilderness boundaries, and intense logging was conducted in the surroundings, rendering the landscape prone to future intense burns and biodiversity loss (DellaSala 2006). Over-reliance on fire suppression and post-fire logging is arguably a shortcoming of fire management strategies in general and the NWFP in particular (DellaSala et al. 2015a). Specifically, the NWFP does not include strict provisions for maintaining LSRs post-intense burn, nor does it focus on viable populations of fire-dependent species and is therefore prone to post-fire biodiversity losses from logging.

Legacy Management

Given the importance of legacies, can management mimic natural landscape patterns and disturbance processes responsible for legacies?

This question has been central to much research, trial-and-error management (often without any trials), and considerable controversy. In the Pacific Northwest, most of the older-aged trees have been logged (Strittholt et al. 2006). Consequently, complex, early-seral forests are now considered a rare type because they are almost always logged following a fire (DellaSala et al. 2015c). However, the closer management can come to mimicking natural processes in form and function, the more likely it can sustain at least some legacy elements that can "lifeboat" important ecosystem functions from the pre- to post-disturbed environment (Hunter 1999). Studies of legacy structures in this region provide a foundation for determining baseline or reference conditions for managers wishing to provide specific structural features, such as snags and down wood, particularly in relation to fire severity (see Sillett and Goslin 1999; Bull 2002; Weisberg 2004; Keeton and Franklin 2005; Tepley et al. 2013; Brown et al. 2013; O'Halloran et al. 2014; Merschel et al. 2014; Dunn and Bailey 2016). Here I build on these recommendations with specific actions for improving upon the NWFP so managers can aptly recognize the ecosystem benefits of fire.

- *Develop clear, measurable objectives that place ecosystem benefits of fire on par with decisions to suppress fire.* An ecosystem benefits analysis (Ingalsbee and Roja 2015; DellaSala et al. 2017; also see www.forestandrangelands.gov) would allow managers to evaluate the conditions under which fires can be managed for such benefits, and determine how best to weigh tradeoffs and collateral damages to ecosystems from pre- and post-fire logging. Assessments should include long-term benefits of fire in creating pyro-diverse landscapes, complex-early seral structures, and fire refugia as part of multiple resource objectives. Ecosystem benefits could then be quantified and integrated with suppression, community safety, and other values (Ingalsbee and Roja 2015; DellaSala et al. 2015c, 2017).
- *Manage large fires for pyrodiversity in protected areas and legacy landscapes on public lands.* The more that fire is managed for pyrodiversity effects, the more the landscape has capacity to generate legacies. To minimize human-safety conflicts, large fires can be compartmentalized so firefighting resources target the fire front closest to homes, whereas other portions of the fire in remote areas are managed for ecosystem benefits (Ingalsbee and Roja 2015).
- *Manage individual forests and landscapes for complexity before disturbance.* Management objectives that include retention of pre-disturbance stand and landscape structures provide the foundation for legacies to regenerate when a natural disturbance eventually happens. Managers may provide ecosystem benefits by altering the quantity, quality, and spatial arrangement of forest structures at stand and landscape scales (Figure 3.8). In particular, restoration of degraded sites should focus on promoting large, irregularly shaped and clumped spacing of large-legacy structures well distributed at the stand and landscape scale, which is preferred over small-homogenized and widely spaced legacies.

Managers can also create structure in otherwise homogeneous plantations by girdling trees in clumps (Figure 3.8C). Irregular spacing of canopy gaps (not shown in Figure 3.8) allows shrubs and forbs to persist by extending the time a site remains in the complex-early stage (Swanson et al. 2011). Also see Zenner (2000), Aubry et al. (2009), and Keeton and Franklin (2005) for effects of varied retention standards on forest ecosystems.

Felling and tipping some large trees into streams where in-stream habitat structure is lacking creates legacies for aquatic organisms (Benda et al. 2016). Large logs left on the ground are "nurse logs" for seedlings and habitat for scores of moisture-dependent invertebrates, fungi, mosses, and herpetofauna (Franklin et al. 2000). "Feathering" edges reduces contrast between managed and unmanaged areas while minimizing edge effects (Hunter 1999).

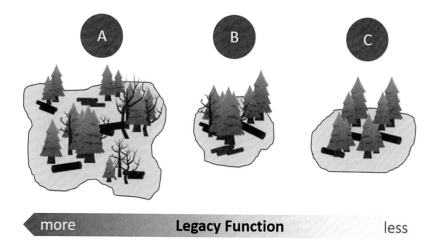

Figure 3.8 Decreasing legacy functions from left to right. Large irregular patches (A) function more as intact legacies, whereas small ones (B) have diminished functions. Clumping of legacies, particularly large snags and logs, is preferred over smaller tree sizes and regular spacing of legacies (C). Restoration involves moving degraded sites (C) to restored (A) sites with intact legacy functions based on reference conditions. (Prepared by C. Mills, Geos Institute.)

- *Manage individual forest stands for legacies after disturbance.* Post-fire logging is a chronic disturbance that can compact soils, remove large legacies (live and dead), retard natural conifer regeneration, and is often coupled with herbicide treatments and road expansion (summarized in Lindenmayer et al. 2008; DellaSala et al. 2015b). In the Pacific Northwest, large-scale post-fire logging is most often proposed even in LSRs, roadless areas, and other "protected areas" based, in part, on the assumption that management is needed to "recover" these areas (DellaSala et al. 2015b). Instead, managers wishing to provide legacies should operate under the premise that if the area was high-quality wildlife habitat before a disturbance (e.g., LSR, roadless area) it remains so after the disturbance because what happens in the post-disturbance environment determines the trajectory of ecosystems through time (i.e., if degraded early in the succession, the subsequent stages may be irreparably harmed; Lindenmayer et al. 2014).
- *Exercise caution in use of back-burns to mediate large fires.* Back-burns are fires intentionally set by managers to slow or reduce the intensity of an advancing fire front by reducing flammable vegetation under more "controlled" conditions before the main fire arrives.

Although important as a fire safety precaution, back-burns often escape containment (e.g., under windy, dry, and hot conditions) and therefore can compound disturbance effects particularly in wilderness and other high-conservation value areas (Backer et al. 2004; Ingalsbee and Roja 2015). Back-burning is often associated with cumulative disturbances, such as bulldozed fire lines that become de facto roads and corridors for flammable weedy plants. Documenting where back-burns have occurred, avoiding additional disturbance (post-fire logging), and restoring fire lines by recontouring slopes and seeding with native plants are critical steps in reducing cumulative impacts. Wherever possible, managers should employ minimal fire suppression tactics (MIST) in roadless areas, wilderness, parks, and other lands of high-conservation value (Ingalsbee and Roja 2015).

- *Restore ecosystems by addressing linked human-caused stressors, particularly roads, livestock, off-highway vehicles, and other chronic disturbances.* The chronic effect of roads on ecosystems is well-documented and generally includes unnatural fire ignitions, spread of invasives, chemical and air pollution, wildlife–vehicle collisions, water-quality degradation, and habitat fragmentation (see Ibisch et al. 2017). Managers may wish to reduce these chronic effects by seasonal road closures (to limit human-caused ignitions), repairing and obliterating failing roads, improving culvert design, and containing flammable weedy plants. The same is true for livestock grazing, off-highway vehicles, and other chronic disturbances. These should be restricted in burn areas. Restoration of areas degraded by intensive management should include both active and passive (remove the stressor) measures that reduce chronic effects.

Conclusions (So What?)

Legacies come in all different shapes, sizes, and functions. They are produced by varied disturbances acting at multiple spatiotemporal scales. Large fires of mixed-severity effects on ecosystems are critical to forest disturbance dynamics because they provide unique pulses of biological legacies not often found in managed forests, except by way of fires that escape containment, nor are they mimicked by low-intensity or prescribed burning (DellaSala and Hanson 2015; DellaSala et al. 2017). The likelihood that legacy landscapes will be managed for ecosystem benefits is increasingly constrained by the economics of forest management and a preoccupation with fire-suppression activities.

Recent research has shown complex, early-seral forests with prolific legacies are as rich in biodiversity as the more-heralded old-growth forests, yet they are greatly underappreciated (Swanson et al. 2011; DellaSala and Hanson 2015; DellaSala et al. 2017). Thus, greater awareness of the importance of these newly emerging forests, the quantity and quality of legacies before and after disturbance (i.e., antecedent conditions and past disturbance legacies), and the large fires that generate them are urgently needed (DellaSala et al. 2015b). In particular, extensive post-fire logging and fire suppression are the main drivers of ecosystem alteration or degradation in fire-adapted systems and why complex, early-seral forests are among the rarest in the Pacific Northwest (Swanson et al. 2011; DellaSala et al. 2015b). Homogenization of landscapes through fire suppression and a desire to manage fires for low-severity vegetation effects may be altering important evolutionary processes that have structured fire-resilient communities for millennia. The emergence of novel climates in conjunction with unprecedented anthropogenic disturbances may flip disturbance-adapted ecosystems to highly altered states that exceed disturbance thresholds (Paine et al. 1998; Lindenmayer et al. 2011). Thus, managers wishing to maintain pyrodiverse landscapes and their associated biota need to manage for complexity before fires, accommodate large fires for ecosystem benefits wherever possible, and limit or prohibit post-fire logging and related activities to allow for the full expression and development of linked seral stages.

Most reserves are too small to fully incorporate large-scale disturbance dynamics unimpeded by fire suppression and thus declines in disturbance-associated biota are anticipated. Reserve-based approaches aptly have focused on late-successional forests in this region because of their conservation status (Strittholt et al. 2006). However, protected areas need to be as large as possible, selected to represent varied patch types in redundant distributions, and the surroundings integrated through compatible management activities that maintain viable populations of well-distributed, disturbance-adapted and disturbance-avoiding species.

We encourage researchers to focus attention on: (1) the unique biota (especially rare and declining species) of these forests and their special adaptations; (2) habitat and population requirements of fire-adapted species; (3) intensity and frequency of fires necessary to maintain fire-adapted species, particularly in a changing climate; (4) spatial distribution, patch types, and sizes needed for these species; and (5) effects of different management intensities on fire-adapted species and ecosystems. Notably, comparative studies are needed that contrast homogenized, early-seral clearcuts, early seral forests with partial legacy retentions, and unaltered complex, early-seral forests (reference conditions). This is necessary to generate relevant legacy stand and landscape prescriptions, aid in restoration of degraded areas, and assuage concerns that fires in these forests are ecologically "catastrophic." Instead, such disturbances are ecologically necessary to maintain pyrodiverse landscapes and the pulse of complex, early-seral structures that follow (DellaSala and Hanson 2015; Lindenmayer et al. 2017).

Acknowledgements

I thank David Lindenmayer and an anonymous reviewer for improvements to an earlier version of this manuscript. Funding for this work was provided to me by the Wilburforce, Weeden, and Leonardo DiCaprio foundations; however, the views expressed herein are exclusively my own.

References

Aubry, K. B., C. B. Halpern, and C. E. Peterson. 2009. Variable-retention harvests in the Pacific Northwest: a review of short-term findings from the DEMO study. *Forest Ecology and Management* 258: 398–408.

Backer, D. M., S. E. Jensen, and G. R. McPherson. 2004. Impacts of fire-suppression activities on natural communities. *Conservation Biology* 18: 937–946.

Baker, W .L. 2012. Implications of spatially extensive historical data from surveys for restoring dry forests of Oregon's eastern Cascades. *Ecosphere* 33(23): 1–39.

Baker, W. 2015. Are high-severity fires burning at much higher rates recently than historically in dry-forest landscapes of the Western USA? *PLoS ONE* 10(9): e0136147. doi:10.1371/journal.pone.0136147.

Benda, L. E., S. E. Litschert, G. Reeves, and R. Pabst. 2016. Thinning and in-stream wood recruitment in riparian second growth forests in coastal Oregon and the use of buffers and tree tipping as mitigation. *Journal of Forest Research* 27(4): 821–836.

Bull, E. 2002. The value of coarse woody debris to vertebrates in the Pacific Northwest. USDA Forest Service General Technical Report PSW-GTR-181.

Bunnell, F. L. 2013. Sustaining cavity-using species: patterns of cavity use and implications to forest management. ISRN Forestry Volume 2013, Article ID 457698, 33 pages. http://dx.doi.org/10.1155/2013/457698.

Brown, M. J., J. Kertis, and M. H. Huff. 2013. Natural tree regeneration and coarse woody debris dynamics after a forest fire in the western Cascade Range. USDA Forestry Service Research Paper PNW-RP-592.

Chen, J., Franklin, J. F, and Spies, T. A. 1992. Vegetation responses to edge environments in old-growth Douglas-fir forests. *Ecological Applications* 2: 87–396.

Dale, V. H., J. F. Franklin, F. J. Swanson, and C. M. Crisafulli. 2005. Ecological responses to the 1980 eruption of Mount St. Helens. In V. H. Dale, F. J. Swanson, and C. M. Crisafulli, editors *Disturbance, Survival, and Succession: Understanding Ecological Responses to the 1980 Eruption of Mount St. Helens*, 3–12. New York: Springer.

DellaSala, D. A. 2006. Fire in the Klamath-Siskiyou ecoregion: protecting and restoring the fire mosaic. In G. Wuerthner, editor, *Wildfire A Century of Failed Forest Policy*, 132–152. Washington, D.C.: Island Press.

DellaSala, D. A., and C. T. Hanson (editors). 2015. *The Ecological Importance of Mixed-severity Fires: Nature's Phoenix*. UK: Elsevier.

DellaSala, D. A., R. Baker, D. Heiken, C. A. Frissell, J. R. Karr, S. K. Nelson, B. R. Noon, D. Olson, and J. Strittholt. 2015a. Building on two decades of ecosystem management and biodiversity conservation under the Northwest Forest Plan, USA. *Forests* 6: 3326–3352; doi:10.3390/f6093326.

DellaSala, D. A., D. B. Lindenmayer, C. T. Hanson, and J. Furnish. 2015b. In the aftermath of fire: logging and related actions degrade mixed- and high-severity burn areas. In D. A. DellaSala and C. T. Hanson, editors, *The Ecological Importance of Mixed-severity Fires: Nature's Phoenix*, 313–347. UK: Elsevier.

DellaSala, D. A., C. T. Hanson, W. L. Baker, R. L. Hutto, R. W. Halsey, D. C. Odion, L. E. Berry, R. W. Abrams, P. Heneberg, and H. Sitters. 2015c. Flight of the phoenix: coexisting with mixed-severity fires. In D. A. DellaSala and C. T. Hanson, editors, *The Ecological Importance of Mixed-severity Fires: Nature's Phoenix*, 372–396. UK: Elsevier.

DellaSala, D. A., R. L. Hutto, C. T. Hanson, M. L. Bond, T. Ingalsbee, D. Odion, and W. L. Baker. 2017. Accommodating mixed-severity fire to restore and maintain ecosystem integrity with a focus on the Sierra Nevada of California, USA. *Fire Ecology* 13: 148–171.

Donato, D. C., J. L. Campbell, and J. F. Franklin. 2012. Multiple successional pathways and precocity in forest development: can some forests be born complex? *Journal of Vegetation Science*: 576–584.

Dunn, C. J., and J. D. Bailey. 2012. Temporal dynamics and decay of coarse wood in early seral habitats of dry-mixed conifer forests in Oregon's eastern Cascades. *Forest Ecology and Management* 276: 71–81.

Fontaine, J. B., D. C. Donato, W. D. Robinson, B. E. Law, and J. B. Kauffman. 2009. Bird communities following high-severity fire: response to single and repeat fires in a mixed-evergreen forest, Oregon, USA. *Forest Ecology and Management* 257: 1496–1504.

Franklin, J. F., D. Lindenmayer, J. M. MacMahon, A. McKee, J. Magnuson, D. A. Perry, R. Waide, and D. Foster. 2000. Threads of continuity. *Conservation Biology in Practice* 1: 9–16.

Franklin, J. F., T. A. Spies, R. Van Pelt, A. B. Carey, D. A. Thornburgh, D. Rae Berg, D. B. Lindenmayer, M. E. Harmon, W. S. Keeton, D. C. Shaw, K. Bible, and J. Chen. 2002. Disturbances and structural development of natural forest ecosystems with silvicultural implications, using Douglas-fir forests as an example. *Forest Ecology and Management* 155: 399–423.

Frelich, L. E. 2008. *Forest Dynamics AND Disturbance Regimes: Studies from Temperate Evergreen-deciduous Forests* (Cambridge Studies in Ecology). Cambridge, UK: Cambridge University Press.

Hanson, C. T., R. L. Sherrif, R. L. Hutto, D. A. DellaSala, T. T. Veblen, and W. L. Baker. 2015. Setting the stage for mixed- and high-severity fire. In D. A. DellaSala and C. T. Hanson, editors, *The Ecological Importance of Mixed-severity Fires: Nature's Phoenix*, 3–22. UK: Elsevier.

Heilman, G. E., Jr., J. R. Strittholt, N. C. Slosser, and D. A. DellaSala. 2002. Forest fragmentation of the conterminous United States: assessing forest intactness through road density and spatial characteristics. *Bioscience* 52: 411–422.

Hessburg, P. F., R. B. Salter, and K. M. James. 2006. Re-examining fire severity relations in pre-management era mixed conifer forests: inferences from landscape patterns of forest structure. *Landscape Ecology* 22: 5–24.

Hunter, M. L. Jr. 1999. *Maintaining Biodiversity in Forest Ecosystems*. Cambridge, UK: Cambridge University Press.

Hutto, R. L. 1995. Composition of bird communities following stand-replacement fires in northern Rocky Mountain (U.S.A.) conifer forests. *Conservation Biology* 9: 1041–1058.

Ibisch, P. L., M. T. Hoffman, S. Kreft, G. Pe'eer, V. Kati, L. Biber-Freudenberger, D. A. DellaSala, M. M. Vale, P. R. Hobson, and N. Selva. 2017. A global map of roadless areas and their conservation status. *Science* 354: 1423–1427.

Ingalsbee, T., and U. Roja. 2015. The rising costs of wildfire suppression and the case for ecological fire use. In D. A. DellaSala and C. T. Hanson, editors, *The Ecological Importance of Mixed-severity Fires: Nature's Phoenix*, 348–371. UK: Elsevier.

Keeton, W. S, and J. F. Franklin. 2005. Do remnant old-growth trees accelerate rates of succession in mature Douglas-fir forests? *Ecological Monographs* 75: 103–118.

Law, B. E., and R. H. Waring. 2015. Carbon implications of current and future effects of drought, fire, and management of Pacific Northwest forests. *Forest Ecology and Management* 355: 4–14.

Lindenmayer, D. B., R. B. Cunningham, and R. Peakall. 2005. The recovery of populations of bush rat *Rattus fuscipes* in forest fragments following major population reduction. *Journal of Applied Ecology* 42: 649–658.

Lindenmayer, D. B., P. J. Burton, and J. F. Franklin. 2008. *Salvage Logging and its Ecological Consequences*. Washington, D.C.: Island Press.

Lindenmayer, D. B., R. J. Hobbs, G. E. Likens, C. J. Krebs, and S. C. Banks. 2011. Newly discovered landscape traps produce regime shifts in wet forests. *PNAS* 108 (38): 15887–15891.

Lindenmayer, D. B., W. Blanchard, L. McBurney, D. Blair, S. C. Banks, D. A. Driscoll, A. L. Smith, and A. M. Gill. 2014. Complex responses of birds to landscape-level fire extent, fire severity and environmental drivers. *Diversity and Distributions* 20: 467–477.

Lindenmayer, D. L., S. Thorn, and S. Banks. 2017. Please do not disturb ecosystems further. *Nature Ecology & Evolution*. doi:10.1038/s41559-016-0031 www.nature.com/natecolevol.

Mackey, B., D. B. Lindenmayer, A. M. Gill, M. A. McCarthy, and J. A. Lindesay. 2002. *Wildlife, Fire and Future Climate: A Forest Ecosystem Analysis*. Melbourne: CSIRO Publishing.

Merschel, A. G., T. A Spies, and E. K Heyerdahl. 2014. Mixed-conifer forests of central Oregon: effects of logging and fire exclusion vary with environment. *Ecological Applications* 24: 1670–1688.

Noss, R. F., and A. Cooperrider. 1994. *Saving Nature's Legacy*. Washington, D.C.: Island Press.

Odion, D. C., J. R. Strittholt, H. Jiang, E. Frost, D. A. DellaSala, and M. Moritz. 2004. Fire severity patterns and forest management in the Klamath National Forest, northwest California, USA. *Conservation Biology* 18: 927–936.

Odion, D. C., C. T. Hanson, A. Arsenault, W. L. Baker, D. A. DellaSala, R. L. Hutto, W. Klenner, M. A. Moritz, R. L. Sherriff, T. T. Veblen, and M. A. Williams. 2014. Examining historical and current mixed-severity fire regimes in ponderosa pine and mixed-conifer forests of western North America. *PLoS One* 9: 1–14.

O'Halloran, T. L, S. A Acker, V. M Joerger, J. Kertis, and B. E. Law. 2014. Postfire influences of snag attrition on albedo and radiative forcing. *Geophysical Research Letters* 41: 9135–9142.

Paine, R. T., M. J. Tegner, and E. A. Johnson. 1998. Compounded perturbations yield ecological surprises. *Ecosystems* 1: 535–545.

Parr, C., and R. Brockett. 1999. Patch-mosaic burning: a new paradigm for fire management in protected areas? *Koedoe* 42: 117–130.

Perry, D. A., P. F. Hessburg, C. N. Skinner, T. A. Spies, S. L. Stephens, A. H. Taylor, J. F. Franklin, B. McComb, and G. Riegel. 2011. The ecology of mixed severity fire regimes in Washington, Oregon, and northern California. *Forest Ecology and Management* 262: 703–717.

Reimchen, T. E. 2000. Some ecological and evolutionary aspects of bear – salmon interactions in coastal British Columbia. *Canadian Journal of Zoology* 78: 448–457.

Schmitz, H., H. Bousack. 2012. Modelling a historic oil-tank fire allows an estimation of the sensitivity of the infrared receptors in Pyrophilous Melanophila beetles. *PLoS One*. http://dx.doi.org/10.1371/journal.pone.0037627.

Sillett, S. C, and M. N Goslin. 1999. Distribution of epiphytic macrolichens in relation to remnant trees in a multi-age Douglas-fir forest. *Canadian Journal of Forest Research* 29: 1204–1215.

Strittholt, J. R., and D. A. DellaSala. 2001. Importance of roadless areas in biodiversity conservation in forested ecosystems: a case study—Klamath-Siskiyou ecoregion, U.S.A. *Conservation Biology* 15: 1742–1754.

Strittholt, J. R., D. A. DellaSala, and H. Jiang. 2006. Status of mature and old-growth forests in the Pacific Northwest, USA. *Conservation Biology* 20: 363–374.

Swanson, M. E., J. F. Franklin, R. L. Beschta, C. M. Crisafulli, D. A. DellaSala, R. L. Hutto, D. B. Lindenmayer, and F. J. Swanson. 2011. The forgotten stage of forest succession: early-successional ecosystems on forested sites. *Frontiers in Ecology and Environment* 9: 117–125. doi:10.1890/090157.

Tepley, A. J, F. J. Swanson, and T. A. Spies. 2013. Fire-mediated pathways of stand development in Douglas-fir/western hemlock forests of the Pacific Northwest, USA. *Ecology* 94: 1729–1743.

Thompson, J. R., T. A. Spies, and L. M. Ganio. 2007. Reburn severity in managed and unmanaged vegetation in a large wildfire. *PNAS* 104(25): 10743–10748.

Weisberg, P. J. 2004. Importance of non-stand-replacing fire for development of forest structure in the Pacific Northwest, USA. *Forest Science* 50: 245–258.

Williams, M. A., and W. L. Baker. 2012. Spatially extensive reconstructions show variable-severity fire and heterogeneous structure in historical western United States dry forests. *Global Ecology and Biogeography* 21: 1042–1052.

Zald, H. S. J., and C. J. Dunn. 2018. Severe fire weather and intensive forest management increase fire severity in a multi-ownership landscape. *Ecological Applications* 28: 1–13.

Zenner, E. K. 2000. Do residual trees increase structural complexity in Pacific Northwest coniferous forests? *Ecological Applications* 10: 800–810.

Emerging Disturbances Amidst Global Change

Non-native Species, Disease, and Synergies with other Disturbances

chapter 4

Context-dependent Effects of Livestock Grazing in Deserts of Western North America

Kari E. Veblen, Erik A. Beever and David A. Pyke

Introduction

Grazing of herbaceous (i.e., non-woody) plants by large ungulates, both native and domestic, occurs throughout the world's rangelands. Large ungulate grazing is a fundamental ecological process in many ecosystems, and can often be sustainably managed. However, inappropriate domestic livestock grazing practices have been implicated in dramatic changes to landscapes across the western United States and elsewhere around the world. Particularly in more climatically severe environments (e.g., desert ecosystems characterized by slow, episodic plant growth), extreme climatic events, such as drought, can interact with historic and contemporary grazing disturbances to exacerbate degradation. These interacting disturbances can lead to plant communities that are less resistant to invasion by undesirable plants species, less resilient to disturbances such as fire or future grazing, and virtually incapable of natural recovery in the absence of active restoration efforts. Management of these ecosystems requires adaptive use of appropriate timing, frequency, and intensity of grazing (Briske 2011; Allen et al. 2017)—or, in some cases, cessation of grazing entirely—to maintain ecosystem health or to recover from grazing-induced degradation. In this chapter, we first provide a general review of grazing disturbance by large mammalian grazers and the role of ecological context in moderating its effects, with emphasis on North American deserts. We then discuss the ecological consequences of cessation of livestock grazing and present a case study from the Mojave Desert, USA.

Grazing as a Disturbance

A primary effect of grazing is selective removal and ingestion of herbaceous plants, in contrast to removal of woody biomass from woody plants by browsing herbivores. Grazing directly influences plants, for example positively by stimulating growth via removal of meristems (Oesterheld and McNaughton 1991; Paige 1999), or negatively by inflicting enough damage to retard growth or cause death (Huntly 1991). Depending on grazer preferences, some plant species (e.g., particularly palatable ones) may decline in relative abundance in the plant community, whereas less-palatable ones increase (Augustine and McNaughton 1998). Likewise, relative abundance of plants with certain life-history strategies, for example, those of low competitive ability or low tolerance to herbivory may increase or decrease, respectively, in response to grazing (Lind et al. 2013). Additionally, the relative abundance of plants with different pho-tosynthetic pathways (e.g., C_3 vs. C_4) may shift in response to the timing of grazing, especially relative to plant phenology and season of greatest vulnerability to defoliation (Lauenroth et al. 1994). Together, responses of plants at the individual and population levels are manifested as broader alterations to plant community composition across landscapes (Huntly 1991). In some cases, this can be positive, such as the reduction of domi-nant plant species to allow more-diverse plant communities. In other cases, alterations may be negative, such as shifting competitive balance away from desirable and toward undesirable (e.g., invasive) species.

Ungulate grazing can have a multitude of effects beyond selective defoliation (Beever et al. 2018). Grazer hoof action can compact soils, reduce soil stability, and reduce water infiltration (Weltz et al. 1989; Greenwood and McKenzie 2001)—effects that may persist for decades (Webb 2002; Castellano and Valone 2007). Grazing animals also can redis-tribute nutrient pools and influence nutrient cycling through consump-tion, urination, and defecation (Augustine 2003, Frank 2008). Additionally, depending on their densities, animals can either promote or reduce both fine- and broad-scale heterogeneity in soil nutrients and vegetation (Adler et al. 2001; Augustine and Frank 2001; Seefeldt and Leytem 2011; Veblen 2012; Koerner et al. 2018). Finally, many ungulate grazers can alter patch dynamics through behaviors such as wallowing (Hobbs 1996); this creates habitat not only for colonization by ruderal plants, but also for smaller animals (e.g., western whiptail lizards, kangaroo mice) that are associ-ated with more-open habitats. Both the magnitude and net positive vs. negative impacts of grazing depend greatly on ecosystem context (see below). In some cases, grazers can be deliberately managed to increase landscape heterogeneity to benefit wildlife (Porensky and Veblen 2015), whereas in other cases grazing is more likely to homogenize the land-scape (Adler et al. 2001). It also may be possible to manage grazers to con-sume and decrease exotic grass (Diamond et al. 2009) or shrub (Dziba et al. 2007) cover and, depending on the properties of the plant community, reduce fire risk by reducing fuel loads (Strand et al. 2014). However, if grazing also promotes invasion of undesirable plant species (for example, invasive species in areas prone to degradation), the costs may outweigh any benefits. Beyond the direct effects of grazing animals on plants and soils (that may cascade to wildlife and other ecosystem components), the broader enterprise of grazing management also can affect wildlife and their habitats. For example, this can occur via "pest" control (e.g., removal of prairie dogs), or building of roads or water points (Freilich et al. 2003).

Context-dependence of Grazing Effects

The consequences of grazing—and resilience of a system to grazing disturbance—are highly context-dependent and vary across rangelands globally. Particularly important is biophysical context, which can supersede the more-obvious effects of grazing intensity or duration (Milchunas and Lauenroth 1993). For example, climatic context strongly influences how an ecosystem responds to grazing disturbance. Plants are better able to respond to grazing when moisture is more readily available and predictable (Milchunas et al. 1988). Rainfall in warmer, drier ecosystems is not only lower but also more variable (Brooks and Chambers 2011); higher temperatures further reduce water availability through increased evaporation. Accordingly, cooler, moister regions tend to be more resilient to disturbance over shorter timescales, whereas warmer, drier regions experience more protracted recovery from disturbance (Chambers et al. 2014a). Extreme stress (e.g., grazing during drought) repeated for multiple years can push plant communities over degradation thresholds from which they are unlikely to recover.

Plant responses to regional climate conditions are further modified by site-level factors, such as landscape position (e.g., potential for run-off water inputs from higher elevations, reduction of water loss due to shading by adjacent mountains) and northern vs. southern aspect. Additional site factors such as geology, soil parent material, and topography can impact soil water-holding capacity, erodibility, and soil nutrients. Local soil conditions also moderate moisture relationships and response to grazing disturbance. For example, finer-textured soils are significantly more likely to experience compaction, especially if grazing occurs when soils are wet (e.g., Van Haveren 1983, Beever and Herrick 2006). Synergistic interactions between soil depth and plant structural properties, such as rooting depth and water-use efficiency, also influence plant access to water, and therefore moderate plant responses to drought (Munson et al. 2015) and resilience to grazing. Additionally, presence of water-holding or restrictive layers in the soil can increase moisture availability for plants (e.g., Duniway et al. 2007).

Evolutionary history further influences plant responses to grazing (Milchunas and Lauenroth 1993). In general, plant responses are influenced by tolerance to, or avoidance of, disturbance (Pyke et al. 2010, 2016), particularly via adaptation to evolution with grazing animals (Mack and

Thompson 1982; Strauss and Agrawal 1999). In these plant communities, resistance to disturbance can be increased by plant morphological characteristics (e.g., short stature), structural mechanisms (e.g., spines or thorns), chemical adaptations (e.g., secondary compounds), or growth characteristics (e.g., clonal growth, association with less palatable plants), all of which deter herbivory (Milchunas and Noy-Meir 2002). Plants and soils generally are better able to sustain grazing in places such as the prairies of the Great Plains, USA or the savannas of Africa, both of which evolved with large herds of grazing ungulates (Mack and Thompson 1982). It has been suggested that, for ecosystems that have evolved with grazing but lost native wild herbivore populations, domestic livestock grazing may be necessary to maintain ecosystem structure and function (Perevolotsky and Seligman 1998; Cingolani et al. 2005). Such suggestions assume that domestic herbivore grazing functionally mimics processes associated with the missing native herbivores, which may only partially be true (Veblen et al. 2016). In contrast, plants in North American cold deserts, such as the Great Basin, or hot deserts, such as the Mojave Desert, are less adapted to grazing because, historically, grazing animals were low in number and typically distributed sparsely throughout these regions (Mack and Thompson 1982).

Disturbance Novelty and Intensity

In some ecosystems, livestock grazing constitutes a novel or intensified disturbance. For example, some areas evolved with mixed feeders (e.g., elk and pronghorn) and browsers (e.g., deer), which together exert substantial herbivory pressure on shrubs and the forb (i.e., non-grass) component of the herbaceous plant community. Cattle, on the other hand, constitute a more-novel disturbance because they are grazers that favor, and therefore increase the intensity of disturbance on, grasses. Additionally, stocking rates of livestock are often higher than historical numbers of native herbivores (Mack and Thompson 1982), thereby creating higher intensity and frequency of disturbance. Such differences in stocking rates are likely a primary distinguishing characteristic between effects on plant communities due to cattle vs. native herbivores (Veblen et al. 2015, 2016). Similarly, duration of grazing may constitute novel or intensified disturbances. Whereas native herbivores often move quickly through an area (at least seasonally, and often within hours to weeks) (e.g., Geremia et al. 2014; Merkle et al. 2016), fenced cattle may stay in a given area longer, depending on the grazing system (e.g., rest-rotation schedules). Consequently, some grazing systems allow cattle to graze plants more intensely and/or more often, which can be injurious to some plant species. Grazing-management systems that employ artificial water enhancements (e.g., stock tanks) also constitute a dramatic departure from pre-historic conditions. Specifically, such enhancements allow livestock producers to facilitate movement of grazing pressure to different parts of the landscape and thus avoid ecosystem degradation caused by the poor distribution of livestock. Additionally, artificial wildlife watering points may be used to decrease mortality of ungulate game species, especially in drier years. But when wildlife water points are not properly fenced and maintained, they also are used by domestic livestock and may lead to unfavorable livestock distributions.

A plant community in its reference state (sensu Bestelmeyer et al. 2017) is expected to shift among different phases (e.g., shifts between grass and shrub dominance in sagebrush steppe; Pyke 2011). However, when disturbance is novel or extreme, the plant community is more likely to permanently cross a threshold to an alternative, often degraded state. For example, many of western North America's rangelands are now more susceptible to degradation due to historic high-intensity grazing. These rangelands, particularly desert ecosystems, were grazed far beyond natural capacity around 1890–1920, which led to widespread, undisputed degradation (Borman 2005; Pyke et al. 2016). Although establishment of the 1934 Taylor Grazing Act led to more-moderate grazing practices, this legacy of historic disturbances interacts with present-day grazing disturbance. Some areas that experienced sustained above-capacity stocking rates 100 years ago still possess degraded soil conditions and a depauperate native plant community. Because these conditions typically reduce resilience to present-day grazing disturbance, even moderate grazing cannot be sustained predictably in these areas, largely due to the presence and temporally variable productivity of invasive plants.

Present-day grazing disturbance also has important consequences for desert ecosystems. Different livestock grazing systems are characterized by specific timing, duration, and intensity of grazing, as well as incorporation of periods of rest from grazing. Manipulating these factors is typically intended to maximize and sustain plant production and, ideally, promote plant recovery from grazing. For example, in parts of the Great Basin Desert where rainfall is comparatively high (e.g., in the Basin's northeast corner), livestock grazing before the activation of intercalary meristematic growth provides adequate opportunity for regrowth and seed production (Briske and Richards 1995). In addition, grazing during the summer dormant season appears to be less harmful because, during this time period, dead plant material is more commonly removed than is new growth (Strand et al. 2014). In contrast, warm-desert ecosystems, where rainfall is lower and more variable, require more flexibility to move livestock when and where precipitation and production dictate. The most predictable grazing season is summer, when plant growth increases with summer monsoonal moisture, but storms may not deliver precipitation equally across time or space. Therefore, moving livestock based on season and times or areas of higher primary productivity often optimizes plant and livestock production (Howery et al. 2000).

Grazing management can be made more complex and challenging by the presence of free-ranging animals. In the western Great Basin, for example, increased numbers of free-ranging horses have been shown to decrease richness and diversity of plants and animals (Beever et al. 2003; Beever and Herrick 2006; Beever et al. 2008). Likewise, in the montane pampas grasslands in Argentina, Loydi and Zalba (2009) found that dung piles of feral horses (which covered 2.5 percent of the study area) acted as "invasion windows" for non-native plants. Analogously, grazing by free-roaming horses in the coastal marshes of the U.S. southeast induced a cascade of ecological changes that spanned numerous trophic levels (Levin et al. 2002). Disturbance by native fauna, such as native ungulates, small mammals, and ants, also can interact with livestock-grazing effects to exacerbate or mitigate disturbance. Conversely, fossorial animals can mitigate negative effects of grazing by aerating compacted soils (Beever et al. 2008).

The challenges of grazing management are further compounded by contemporary climate change. The deserts of western North America have already been experiencing elevated temperatures and more-frequent occurrence of extreme conditions such as heat waves or prolonged droughts (Mote and Redmond 2011). These conditions are projected to intensify (Polley et al. 2013; IPCC 2018), and ecosystems will experience novel and intensified levels of climate-driven disturbance. Deserts of the western U.S. generally are expected to experience higher temperatures and drier conditions (Polley et al. 2013). As discussed above, reduced rainfall coupled with higher temperatures has direct implications for resilience to grazing disturbance, loss of perennial vegetation, and rate of degradation. Moreover, livestock forage under the projected future climate is likely to be less-abundant and less-reliable, making grazing management even more challenging.

Grazing Disturbance and the Intermediate Disturbance Hypothesis

Application of the Intermediate Disturbance Hypothesis (IDH) to grazing disturbance has been relatively infrequently tested relative to other ecological disturbances. In research on free-roaming horses, Beever et al. (2008) found that across three indices of grazing disturbance and two measures of plant species richness across the western Great Basin (USA), ecosystem response to disturbance was better described by a negative linear relationship than a quadratic (bell-shaped) relationship indicative of IDH. That is, as levels of grazing disturbance increased, species richness monotonically declined. Strikingly, these patterns were also observed in measures of small-mammal richness and level of soil compaction, but not necessarily in measures of ant-mound abundance (Beever 1999). In contrast, support for the IDH was found on the northern slope of Mt. Kilimanjaro in Tanzania, an area in which plants have co-evolved with numerous species of ungulates; plant species diversity was higher on moderately and lightly livestock-grazed areas than in areas with heavy or no grazing (Kikoti and Mligo 2015). In a third and final illustrative example, in northeastern China meadow-steppe plant communities, Yan et al. (2015) found that plant species richness was highest at an intermediate stocking rate, but only in the third and final year of the study, whereas standing crop of canopy biomass was highest at intermediate stocking rates in all three years. Hobbs and Huenneke (1992) provide a thoughtful review of IDH for grazing and several other ecological disturbances. They note that many studies have reported highest species diversity under intermediate levels of grazing, and that in some areas, such as Mediterranean-climate grasslands and the chalk grasslands of northern Europe, cessation of grazing leads to dominance by a few grass species, encroachment by woody species, and increased abundance of non-native species.

Shrub Encroachment and Invasive Species

Changes in the structure of desert plant communities are commonly associated with the virtually permanent crossing of thresholds to alternative ecosystem states. In deserts, such changes often take two forms. The first is woody plant encroachment, a shift from continuous cover of either perennial grasses or grass–shrub mixtures to discontinuous vegetation dominated by trees or shrubs that would typically be at low abundance on a site (Archer et al. 2017). The resulting woody-dominated communities typically are characterized by lower annual productivity, have biomass primarily stored in woody material, and have exposed soils in interspaces between perennial vegetation. The second form begins as a discontinuous, shrub-dominated or mixed grass–shrub system that supports an interspersion of perennial plants with biological soil crusts or desert pavement. This system then shifts to a continuous cover of herbaceous plants often dominated by invasive (often annual) grasses; the resulting plant community lacks both diversity and the native-perennial plant component that serves as livestock forage and habitat for many wildlife species.

These types of shifts in vegetation states are driven by stressors that may operate and interact at multiple spatial and temporal scales simultaneously (Peterson et al. 1998; Walker et al. 2002). Over geologic time, woody plant dominance in the western dryland ecosystems of North America has ebbed and flowed, depending on climate fluctuations (Van Devender 1995). Most recently, during the warming of the Holocene, the dominance of trees was reduced in favor of the grasses and shrubs that dominate most western U.S. deserts today (Miller and Wigand 1994; Van Devender 1995). Vegetation states also reflect climate over broad spatial gradients. Regionally, areas typically having warm, wet growing seasons are more resilient to disturbance (and less likely to cross thresholds to shrub- or invasive-dominated states) than are areas with cooler and drier growing seasons. Climate projections for the next century suggest that increased temperatures and variability in precipitation will reduce productivity of desert rangelands of the western U.S. and increase vulnerability to stressors, such as invasive species (Polley et al. 2013).

At the landscape scale, woody-plant encroachment and exotic-plant invasions in western deserts occur due to several interrelated causes. A common contributor to both phenomena is inappropriate livestock-grazing practice, in particular, over-utilization of vegetation and inappropriate timing of grazing. For example, in black grama grasslands of the Chihuahuan Desert, inappropriate livestock grazing during drought can reduce perennial vegetation and create bare soil patches. Wind erosion can then distribute soil under mesquite shrubs to form dunes and shift the landscape from grassland to shrubland (Peters and Gibbens 2006; Bestelmeyer et al. 2015). Similarly, a combination of inappropriate livestock-grazing practices, fire control, and elevated levels of CO_2 have increased juniper and pinyon pine in the Great Basin and Colorado Plateau over the last 150 years (Van Auken 2009). Invasive grasses also tend to increase in response to livestock-driven reductions in cover of native perennial grass and biological soil crusts (Reisner et al. 2013).

Although, historically, climate and fire often dictated changes in plant dominance, anthropogenic disturbances—in particular, livestock grazing and introduced plant species—now interact with these factors to shape desert plant communities. The Sonoran and Chihuahuan Desert grasslands and the cooler and moister portions of the Great Basin likely had fire-return frequencies as short as ≤25 years (McPherson 1995; Peters and Gibbens 2006; Miller and Heyerdahl 2008), whereas warmer and drier sites of the Great Basin and Mojave deserts likely had much longer intervals between fires, likely exceeding 50 years (Brooks 1999; Miller et al. 2013). In the cooler and moister areas of the Great Basin, livestock overuse of the herbaceous vegetation in the late 1800s through the 1950s limited the extent and frequency of fires within these ecosystems and contributed to the increase in cover of piñon pine and junipers. In the Sonoran Desert, perennial grasses from Africa and Asia, such as Lehman lovegrass (*Eragrostis lehmanniana*) and buffelgrass (*Pennisetum ciliare*), were intentionally introduced in the 1930s to provide livestock forage (McPherson 1995). In the warmer and drier portions of the Great Basin and Mojave Deserts, Euro-American settlement led to unintentional introductions (via crop-seed contamination) of non-native annual grasses from Eurasia such as cheatgrass (*Bromus tectorum*) and red brome (*B. rubens*) (Pyke et al. 2016). All of these introduced species have become invasive, are generally considered undesirable, and are further promoted by inappropriate livestock grazing practices. Increasing dominance by these species has led to positive feedbacks whereby increased fine (grass) fuels promote greater fire extent and more-frequent fires (Anable et al. 1992; D'Antonio and Vitousek 1992). In turn, these promote further plant invasions and reduce cover of desirable perennial plants.

Human population growth is likely to exacerbate current problems associated with invasive species. Urban growth within desert regions is increasing at rates higher than the national average (Albrecht 2008) and is accompanied by increased use of fossil fuels and local to regional changes in atmospheric gases. Increases in atmospheric concentrations of CO_2 and N will likely contribute to increases in invasive grasses (Nowak et al. 2004; Smith et al. 2014), particularly near cities (Galloway et al. 2003). Similarly, transportation and energy corridors (that now cover nearly 1.25 percent of the land area in the western USA; Leu et al. 2008) contribute to both the spread of invasive grasses and increased risk of fires (Gelbard and Belnap 2003; Gelbard and Harrison 2003; Bradley 2010; Parisien et al. 2012).

Removal of Grazing Disturbance

Intuitively, one might expect that removing a disturbance would allow a system to return to its pre-disturbance state. Consequently, conservation and restoration recommendations in rangelands often include removal of livestock grazing for the purposes of ecosystem "recovery". Recovery may in some cases entail return to a pre-disturbance state, but in other cases may entail the recovery of key functions, e.g., stabilized soils, appropriate hydrology, or maintenance of plant cover. However, a key determinant of whether or not removal of grazing alone will lead to ecosystem recovery involves whether or not biophysical thresholds have been crossed. Examples of such thresholds include: (a) loss of local seed sources for plant species that typically decline under high herbivory pressure; (b) loss of enough topsoil that functionality of surface horizons is compromised; (c) downcutting of streambanks that leads to vertical decoupling of the stream channel from the historic floodplain and water table; (d) compaction of soil surfaces and loss of micropore space to the point that incident rain no longer penetrates the surface; and (e) alteration of soil chemical or physical structure due to invasion by non-native species. In North American deserts, some of these thresholds may have been crossed long ago, leaving grazing "legacies" that complicate the ability to anticipate a system's response to removal of contemporary grazing disturbance (Borman 2005). Often, crossing of these thresholds and the speed of recovery are determined by grazing intensity prior to grazing removal

(Liu et al. 2011). For example, in the Mojave Desert, areas that have been altered the most by prior grazing are the most susceptible to invasions by non-native plants in heavy precipitation years following grazing removal (Tagestad et al. 2016).

Ecosystem responses to grazing removal also reflect the timelines of the specific physical and biotic components involved. For example, biological soil crusts are considered bio-indicators of ecosystem health that are generally adversely affected by grazing (Warren and Eldridge 2001) and may require many years to several decades to regenerate following release from grazing (Anderson et al. 1982; Concostrina-Zubiri et al. 2014). Similarly, slow-growing, long-lived plants with long generation times may take a long time to respond. On the other hand, small, ruderal plants may quickly become numerous after grazing is removed and may stabilize soils. Responses of plants can be decoupled from responses of soils. For example, McGovern et al. (2014) found that removal of sheep grazing in montane grasslands resulted in increasing shrub dominance, an indicator of ecosystem recovery. However, soils also showed higher levels of nitrogen-fixing bacteria, soil nitrates, and acidification, indicative of negative effects of grazing removal.

Landscape context and the spatial extent of disturbance influence the short- and long-term ecosystem consequences of that disturbance (Brown and Allen 1989). For example, whereas small ($4\,m^2$) wallowing areas created by free-roaming horses may mimic historic bison wallows or resting areas for native cervids, grazer removal of vegetation cover in areas adjacent to the wallows can interact with broader scale patterns and processes of degradation, such as those that led to the Dust Bowl of the 1930s (Peters et al. 2004). When disturbance is widespread and severe, recovery of key ecosystem processes, such as dispersal and recolonization, may not occur following grazing removal alone and may require active restoration efforts. Similarly, any land-use activities that change land connectivity (e.g., agriculture, rural housing and infrastructure, oil and gas development) may inhibit recovery.

A critical factor in the trajectory of response to grazing removal is potential synergism with other disturbances, both natural and anthropogenic. The potential of a plant community to respond to removal of a single stressor (e.g., grazing) by returning to its original community likely depends on the specific state of the plant community and presence of other stressors (Bestelmeyer et al. 2015). Often, modifying only one of several land-use practices may only partially halt degradation, and, depending on the strength of the stressor, may require additional time to realize recovery or improvements. In desert ecosystems, disturbances that can inhibit responses of soils and desirable plants to release from grazing include uncharacteristically severe or frequent fires, invasive plant or animal species, and herbivory by other animals (e.g., lagomorphs or grasshoppers).

Severe and prolonged drought (Hunt 2001) and other climatic factors, along with the inherent productivity of the system, are important predictors of how a system will respond to livestock removal. Areas with low productivity and low, unpredictable precipitation, such as the Mojave Desert, may recover very slowly from disturbance (Lovich and Bainbridge 1999). However, at the same time, these areas can be relatively resistant to invasion of non-native plants, because few non-natives can establish and persist in these exceedingly harsh environments (Brooks 2009). Recovery trajectories of communities with higher productivity and more-predictable moisture, on the other hand, tend to be more rapid. However, even in very productive areas, extreme degradation can inhibit recovery. For example, riparian areas can be some of the fastest areas to recover in water-limited landscapes (e.g., Elmore and Kauffman 1994; Trimble and Mendel 1995) due to water availability and upstream nutrient inputs. However, riparian areas also can be the slowest to recover because they often are some of the most heavily used parts of a landscape by cattle (Kauffman and Krueger 1984; Trimble and Mendel 1995). The interactive role of drought and grazing will only become more important as regional trends in climate are predicted to include increased temperatures and less precipitation in the U.S. desert southwest.

Bird and Small-mammal Responses

Grazing removal often is intended to promote recovery of animal communities and typically is assumed to follow or coincide with recovery of plants and soils. Given their diverse life-histories, birds and small mammals can serve as "ecological integrators" that provide broader-ecosystem insights regarding the consequences of grazing or its removal. In a meta-analysis across studies that occurred in shortgrass-steppe communities, Milchunas et al. (1988) found that diversity, abundance, dominance, and dissimilarity responses to grazing varied greatly across classes of organisms. In some cases, even though plant-community responses to grazing were relatively minor, animal responses were large (Milchunas et al. 1988).

Areas with long-term grazing removal typically exhibit higher abundance, evenness, and species richness of birds, compared with grazed areas (e.g., Schulz and Leininger 1991; Popotnik and Giuliano 2000; Tewksbury et al. 2002). Grazing effects on birds can include nest trampling, direct disturbance to birds, and indirect effects on bird species' habitat, foraging, and predators (Powers and Glimp 1996; Levin et al. 2002; Beever and Aldridge 2011; Monroe et al. 2017). Abundance and diversity responses of birds to grazing removal can be slow (e.g., Desmond 2004), particularly if vegetation responses are slow (e.g., in deserts). Similarly, bird-community responses to grazing removal may not be as strong as for other taxa, such as plants or small mammals (Moser 1997). However, particularly in wetter areas, birds can respond positively soon after grazing removal. For example, Dobkin et al. (1998) compared the structure and trends of plant and avian communities in response to short-term (4-year) vs. long-term (>30-year) removal from livestock grazing and found higher bird abundance and species richness under long-term removal. Although species composition differed between short- and long-term plots, birds characteristic of the long-term plots began appearing in the short-term plots by the third and fourth years (Dobkin et al. 1998). Similarly, in a nine-year study in Elko County, Nevada, riparian-obligate birds remained absent from grazed sites, but in grazing-removal areas, bird abundance and species richness fluctuated for the first six years and then increased significantly (Bradley et al. 2000).

For small mammals, the effects of grazing (and its removal) may be indirect (e.g., through vegetative cover; Grant et al. 1982) or direct (e.g., through competition with large grazers for food). Small-mammal responses to livestock removal appear to be highly context-dependent, varying by small-mammal species, productivity of the ecosystem, degree of habitat disturbance due to grazing (Grant et al. 1982), and the effects of the grazing regime (particularly stocking rate and season of grazing) on soils and plant cover and diversity (Hanley and Page 1981). Grazing removal often (but not always) promotes less-common small mammal species at the expense of more-common, often ruderal species, although lags can sometimes be pronounced. For example, in western North America, the deer mouse (*Peromyscus maniculatus*) performs well in grazed areas because it is eurytopic and has a ruderal life-history strategy (e.g., Black and Frischknecht 1971; Schulz and Leininger 1991; Bradley et al. 2000; Beever and Brussard 2004). In sagebrush-dominated sites in the Great Basin Desert, *P. maniculatus* decreases in response to removal of heavy grazing by free-roaming horses, whereas "completeness" of the small-mammal community increases (Beever and Brussard 2004). In riparian areas, grazing removal also can lead to long-term (26-year) shifts in rodent communities characterized by decreases in *P. maniculatus* and increases in other small mammal species, such as jumping mice and montane shrews (Schulz and Leininger 1991). Bradley et al. (2000) documented community shifts in riparian areas within five years after grazing removal in which *P. maniculatus* decreased, whereas mountain voles (*Microtus montanus*), western jumping mice (*Zapus princeps*), and vagrant shrews (*Sorex vagrans*) increased.

Mojave Case Study

To illustrate the importance of disturbance intensity, time, and spatial scale as factors in ecosystem recovery from grazing, we present here a case study from the Mojave Desert. In this study, livestock and free-ranging burros were removed from the study area to aid habitat conservation for the federally threatened desert tortoise (*Gopherus agassizii*) (Oldemeyer 1994; Avery and Neibergs 1997). The Mojave Desert is a climatically severe ecosystem (annual precipitation in grazed areas averages ~10–20 cm) that historically had low densities of native grazers. Nonetheless, much of the Mojave Desert has long been grazed by domestic livestock, and in many places exotic annual grasses have supplanted native plant communities, altered fire regimes, and proven very difficult to eradicate (Beatley 1966; Lovich and Bainbridge 1999; Brooks et al. 2004; Brooks and Matchett 2006). Livestock were grazed in the area now encompassed by Mojave National Preserve (MNP) since the 1860s, but beginning in late 2001, they were gradually removed in an attempt to improve tortoise habitat.

We investigated the effects of historic grazing intensity and time since grazer removal on plant and soil attributes in the two dominant plant communities of MNP: white bursage—creosote bush (*Ambrosia dumosa*—*Larrea tridentata*), and Joshua tree—blackbrush (*Yucca brevifolia*—*Coleogyne ramosissima*). Because cattle activity decreases with increasing distance from water points (Andrew 1988)—especially in dry, hot (maximum temperatures of >50°C), ecosystems such as the Mojave—we sampled at 100 m from water points (high grazing intensity), 400 m (medium intensity) and 1600 m (low intensity). Grazing was removed from five water point sites in 2001 and from the remaining sites in 2002. We assessed the effects of time since grazing removal on plants and soils by sampling plots in previously grazed water point sites 1–2 and 7–8 years after grazing removal, as well as plots inside long-term (40+ years) exclosures. Sampling occurred in a hierarchically nested fashion: across 10 water points that spanned two livestock-grazing allotments, at three distances from each water point, at three points per distance, and along 2–3 transects per point (see Beever et al. 2006).

We found that in heavily grazed areas, grazing removal improved soil stability within 1–2 years of removal (Beever et al. 2006) and increased cover of native grasses (Beever and Pyke 2005). Cover of an invasive annual grass, *Schismus* spp., increased immediately following grazing removal, most clearly under the (previously) highest grazing intensity (Beever et al. 2006). However, by 7–8 years following grazing removal, *Schismus* decreased, whereas a different annual invasive, *Erodium cicutarium*, increased (Figure 4.1). *Erodium cicutarium* is an increasingly common invader in the Mojave, Sonoran, and Chihuahuan Deserts (Brooks 2000; Allington et al. 2012; Kimball et al. 2014). Importantly, results of our multi-scale sampling illustrated that not only pre-removal gradients, but also patterns in post-removal trajectories, occurred at different spatial resolutions for different plant and soil response variables. Consequently, sampling at the "incorrect" scale(s) would have failed to detect ecosystem disturbance and recovery.

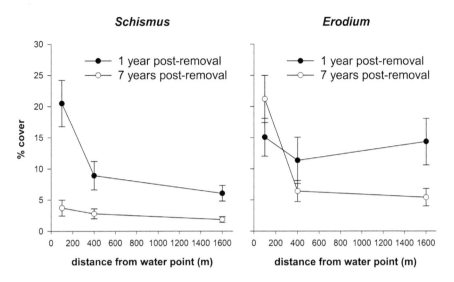

Figure 4.1 Means ±1 SE of cover of two species of annual invasive plants, *Schismus* spp. and *Erodium cicutarium*, at three distances (100 m, 400 m, and 1600 m) from water points. Data were collected from five water points during the wet season in 2003 (1-year post-livestock removal) and 2009 (7-years post-livestock removal). Percentage cover assessed with line-point-intercept sampling (Herrick et al. 2009).

Lower densities of woody plants at 100 m and 400 m relative to 1600 m across the MNP indicated that historic high grazing intensity depressed woody plant densities. These effects persisted from 1–2 years after grazing removal (2003) until 7–8 years following grazing removal (2009) (Table 4.1). Effect-size analyses suggested that, under moderate grazing intensity (400 m from water), densities of both small and larger shrubs were higher under 40 years of protection from livestock than under 1–2 or 7–8 years of protection (Figure 4.2, Table 4.1). Under the highest level of previous grazing intensity (100 m), however, densities of large (maximum height >10 cm) shrubs were still depressed after 40+ years of grazing removal; densities were no higher 40+ years after grazing removal than 1–2 or 7–8 years after removal (Figure 4.2, Table 4.1). Small shrubs (<10 cm height) under high grazing intensity, how-ever, showed signs of recovery by 7–8 years following grazing removal (Figure 4.2, Table 4.1).

Overall, these results illustrate that vegetation recovery from graz-ing can be slow, and suggest that active restoration may be necessary to speed the rate of recovery to meet objectives related to wildlife habi-tat. These two conclusions are especially true in arid ecosystems that have naturally low productivity and only episodic plant germination.

The variation in weather conducive to restoration may require repeated restoration actions to achieve success, similar to recommendations for restoration in the warm and dry portions of the Great Basin (Chambers et al. 2014b).

Table 4.1 Mean and 1 SE of densities of two height classes (<10 cm and >10 cm) of woody plants at three distances (100 m, 400 m, and 1600 m) from water points 1–2 years following grazing removal (2003), 7–8 years following grazing removal (2009), and ≥40 years following grazing removal (long-term exclosures sampled in 2009). Densities of all woody plants were assessed in four to six 50 m × 2 m belt transects at each distance, at each of three sites in the Mojave National Preserve, California, USA

	Distance from well (m)	1–2 years Mean (#/ha) ±SE	7–8 years Mean (#/ha) ±SE	40 years Mean (#/ha) ±SE
<10 cm	100	125 ± 63	283 ± 71	400 ± 87
	400	319 ± 162	278 ± 127	483 ± 169
	1600	811 ± 408	700 ± 367	–
>10 cm	100	5025 ± 685	5492 ± 1146	5750 ± 681
	400	5786 ± 653	5133 ± 623	7917 ± 672
	1600	8622 ± 751	8036 ± 823	–

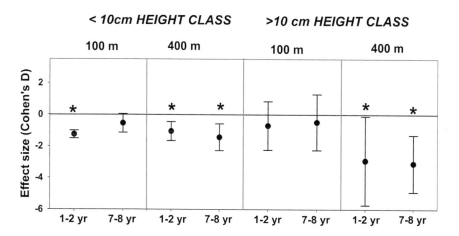

Figure 4.2 Effect size analysis testing woody plant densities 1–2 years and 7–8 years following grazing removal relative to 40+ years following removal (zero line). Data were collected for two height classes (<10 cm and >10 cm) and at two distances (100 m and 400 m) from water points. Points are mean Cohen's D statistic for three sites. Asterisks indicate 90 percent confidence intervals that do not overlap zero.

4.7 Conclusions and Implications

Grazing by domestic ungulates has occurred for millennia in association with human societies. In some cases, introduction of grazing to a system may constitute a novel disturbance that more commonly leads to deterioration, whereas in other systems (especially those with long co-evolutionary histories of grazing), *removal* of grazing constitutes the more-disruptive disturbance. Hobbs and Huenneke (1992) suggested that "... in any situation, a significant change in grazing regime will constitute a disturbance." Aside from the more-predictable influence of evolutionary grazing history, research worldwide collectively has shown that responses to grazing disturbance (and its removal) are highly context-dependent and nuanced. Grazing effects depend strongly on factors, such as the type, season, intensity and duration of grazing; climate; landscape context; and the chemical, physical, and life-history characteristics of individual plant species (Diaz et al. 2007). Furthermore, ungulate grazing can interact with other ecological disturbances—many of which are anthropogenically-driven—such as extreme-weather events (e.g., drought, extended heat waves), wildfire, invasive species, and habitat fragmentation, to affect system response to grazing (e.g., Hobbs 2001; Fuhlendorf et al. 2009).

The nuanced nature of ecosystem responses to grazing challenges the ability of researchers or natural-resource managers to make sweeping generalizations about expected ecological consequences. Therefore, the ability to anticipate the effects of changes to grazing regimes (e.g., increases/decreases in intensity) requires context-specific monitoring of grazing effects. As illustrated both in our Mojave case study and in many other studies, heterogeneity in grazing exists at multiple spatial resolutions. Consequently, the impacts of grazing should be monitored not only at the scale of immediate interest (e.g., plant-community responses to selective foraging), but at both finer (e.g., plant-population) and coarser (e.g., landscape-cover) resolutions to understand the mechanisms behind grazing effects and how they may be manifested at broader scales. Also required will be an evaluation of whether grazing disturbance is likely to promote or impede the achievement of management objectives. Natural disturbances are considered key processes to maintain biodiversity in many systems, but the extent to which livestock grazing may complement, replace, or be compatible with those disturbances will depend heavily on ecological context, the specific grazing practices, and interactions with novel anthropogenic disturbances. Likewise, evaluation of the potential for restoration success in areas that have been degraded by livestock grazing must account for the extent of departure from desired conditions (and therefore likelihood of success) of potential restoration sites.

References

Adler, P. B., D. A. Raff, and W. K. Lauenroth. 2001. The effect of grazing on the spatial heterogeneity of vegetation. *Oecologia* 128: 465–479.

Albrecht, D. E. 2008. *Population Brief: The Changing West – A Regional Overview.* Logan, UT: Western Rural Development Center, Utah State University, pp. 4.

Allen, C. R., D. G. Angeler, J. J. Fontaine, A. S. Garmestani, N. M. Hart, K. L. Pope, and D. Twidwell. 2017. Adaptive management of rangeland systems. In D. D. Briske, editor, *Rangeland Systems: Processes, Management and Challenges,* 373–394. Cham: Springer International.

Allington, G. R. H., D. N. Koons, S. K. Morgan Ernest, M. R. Schutzenhofer, and T. J. Valone. 2012. Niche opportunities and invasion dynamics in a desert annual community. *Ecology Letters* 16: 158–166.

Anable, M. E., M. P. McClaran, and G. B. Ruyle. 1992. Spread of introduced Lehmann lovegrass *Eragrostis lehmanniana* Nees. in southern Arizona, USA. *Biological Conservation* 61: 181–188.

Anderson, D. C., K. T. Harper, and S. R. Rushforth. 1982. Recovery of cryptogamic crusts from grazing on Utah winter ranges. *Journal of Range Management* 35: 355–359.

Andrew, M. H. 1988. Grazing impact in relation to livestock watering points. *Trends in Ecology & Evolution* 3: 336–339.

Archer, S. R., E. M. Andersen, K. I. Predick, S. Schwinning, R. J. Steidl, and S. R. Woods. 2017. Woody plant encroachment: causes and consequences. In D. D. Briske, editor, *Rangeland Systems: Processes, Management and Challenges,* 25–84. Cham: Springer International.

Augustine, D. J. 2003. Long-term, livestock-mediated redistribution of nitrogen and phosphorus in an East African savanna. *Journal of Applied Ecology* 40: 137–149.

Augustine, D. J. and D. A. Frank. 2001. Effects of migratory grazers on spatial heterogeneity of soil nitrogen properties in a grassland ecosystem. *Ecology* 82: 3149–3162.

Augustine, D. J. and S. J. McNaughton. 1998. Ungulate effects on the functional species composition of plant communities: herbivore selectivity and plant tolerance. *Journal of Wildlife Management* 62: 1165–1183.

Avery, H. W. and A. G. Neibergs. 1997. *Effects of Cattle Grazing on the Desert Tortoise, Gopherus agassizii: Nutritional and Behavioral Interactions.* New York: New York Turtle and Tortoise Society & WCS Turtle Recovery Program.

Beatley, J. C. 1966. Ecological status of introduced brome grasses (*Bromus* spp.) in desert vegetation of southern Nevada. *Ecology* 47: 548–554.

Beever, E. A. 1999. Species- and community-level responses to disturbance imposed by feral horse grazing and other management practices. Ph.D. dissertation, University of Nevada, Reno.

Beever, E. A. and C. L. Aldridge. 2011. Influences of free-roaming equids on sagebrush ecosystems, with a focus on Greater Sage-grouse. *Studies in Avian Biology* 38: 273–290.

Beever, E. A. and P. F. Brussard. 2004. Community- and landscape-level responses of reptiles and small mammals to feral-horse grazing in the Great Basin. *Journal of Arid Environments* 59: 271–297.

Beever, E. A. and J. E. Herrick. 2006. Effects of feral horses in Great Basin land-scapes on soils and ants: direct and indirect mechanisms. *Journal of Arid Environments* 66: 96–112.

Beever, E. A., L. Huntsinger, and S. L. Petersen. 2018. Conservation challenges emerging from free-roaming horse management: a vexing social-ecological mismatch. *Biological Conservation* 226: 321–328.

Beever, E. A., M. Huso, and D. A. Pyke. 2006. Multiscale responses of soil stability and invasive plants to removal of non-native grazers from an arid conserva-tion reserve. *Diversity and Distributions* 12: 258–268.

Beever, E. A. and D. A. Pyke. 2005. Short-term responses of desert soil and vegeta-tion to removal of feral burros and domestic cattle. *Ecological Restoration* 23: 279–280.

Beever, E. A., R. J. Tausch, and P. F. Brussard. 2003. Characterizing grazing dis-turbance in semiarid ecosystems across broad scales, using diverse indices. *Ecological Applications* 13: 119–136.

Beever, E. A., R. J. Tausch, and W. E. Thogmartin. 2008. Multi-scale responses of vegetation to removal of horse grazing from Great Basin (USA) mountain ranges. *Plant Ecology* 196: 163–184.

Bestelmeyer, B. T., A. Ash, J. R. Brown, B. Densambuu, M. Fernández-Giménez, J. Johanson, M. Levi, D. Lopez, R. Peinetti, L. Rumpff, and P. Shaver. 2017. State and transition models: theory, applications, and challenges. In D. D. Briske, editor, *Rangeland Systems: Processes, Management and Challenges*, 303–345. Cham: Springer International.

Bestelmeyer, B. T., G. S. Okin, M. C. Duniway, S. R. Archer, N. F. Sayre, J. C. Williamson, and J. E. Herrick. 2015. Desertification, land use, and the trans-formation of global drylands. *Frontiers in Ecology and the Environment* 13: 28–36.

Black, H. L. and N. C. Frischknecht. 1971. *Relative Abundance of Mice on Seeded Sagebrush-Grass Range in Relation to Grazing*, 1–8. USDA Forest Service.

Borman, M. M. 2005. Forest stand dynamics and livestock grazing in historical context. *Conservation Biology* 19: 1658–1662.

Bradley, B. A. 2010. Assessing ecosystem threats from global and regional change: hierarchical modeling of risk to sagebrush ecosystems from climate change, land use and invasive species in Nevada, USA. *Ecography* 33: 198–208.

Bradley, P. V., M. A. Ports, and K. Undlin. 2000. *Wildlife Population Trends in Grazed Riparian Habitats of Northeastern Nevada: A Comparative Study of Healing vs. Ailing Public Lands*. Reno, NV: Nevada Division of Wildlife.

Briske, D. 2011. *Conservation Benefits of Rangeland Practices: Assessment, Recommendations, and Knowledge Gaps*. Washington, D.C.: US Department of Agriculture. Natural Resources Conservation Service.

Briske, D. D. and J. H. Richards. 1995. Plant responses to defoliation: a physio-logical, morphological, and demographic evaluation. In D. J. Bedunah and R. E. Sosebee, editors, *Wildland Plants: Physiological Ecology and Developmental Morphology*, 635–710. Denver, CO: Society for Range Management.

Brooks, M. L. 1999. Alien annual grasses and fire in the Mojave Desert. *Madroño* 46: 13–19.

Brooks, M. L. 2000. Competition between alien annual grasses and native annual plants in the Mojave Desert. *American Midland Naturalist* 144: 92–108.

Brooks, M. L. 2009. Spatial and temporal distribution of non-native plants in upland areas of the Mojave Desert. In R. H. Webb, L. F. Fenstermaker, J. S. Heaton, D. L. Hughson, E. V. McDonald, and D. M. Miller, editors, *The Mojave Desert: Ecosystem Processes and Sustainability*, 101–124. Reno, NV, USA: University of Nevada Press.

Brooks, M. L. and J. C. Chambers. 2011. Resistance to invasion and resilience to fire in desert shrublands of North America. *Rangeland Ecology & Management* 64: 431–438.

Brooks, M. L., C. M. D'Antonio, D. M. Richardson, J. B. Grace, J. E. Keeley, J. M. DiTomaso, R. J. Hobbs, M. Pellant, and D. Pyke. 2004. Effects of invasive alien plants on fire regimes. *Bioscience* 54: 677–688.

Brooks, M. L. and J. R. Matchett. 2006. Spatial and temporal patterns of wildfires in the Mojave Desert, 1980–2004. *Journal of Arid Environments* 67: 148–164.

Brown, B. J. and T. F. H. Allen. 1989. The importance of scale in evaluating herbivory impacts. *Oikos* 54: 189–194.

Castellano, M. J. and T. J. Valone. 2007. Livestock, soil compaction and water infiltration rate: evaluating a potential desertification recovery mechanism. *Journal of Arid Environments* 71: 97–108.

Chambers, J. C., R. F. Miller, D. I. Board, D. A. Pyke, B. A. Roundy, J. B. Grace, E. W. Schupp, and R. J. Tausch. 2014a. Resilience and resistance of sagebrush ecosystems: implications for state and transition models and management treatments. *Rangeland Ecology & Management* 67: 440–454.

Chambers, J. C., D. A. Pyke, J. D. Maestas, M. Pellant, C. S. Boyd, S. B. Campbell, S. Espinosa, D. W. Havlina, E. E. Mayer, and A. Wuenschel. 2014b. Using resistance and resilience concepts to reduce impacts of invasive annual grasses and altered fire regimes on the sagebrush ecosystem and greater sage-grouse: a strategic multi-scale approach. General Technical Report RMRS-GTR-326. Fort Collins, CO: U.S. Department of Agriculture, Forest Service, Rocky Mountain Research Station. 73pp.

Cingolani, A. M., I. Noy-Meir, and S. Diaz. 2005. Grazing effects on rangeland diversity: a synthesis of contemporary models. *Ecological Applications* 15: 757–773.

Concostrina-Zubiri, L., E. Huber-Sannwald, I. Martinez, J. L. F. Flores, J. A. Reyes-Aguero, A. Escudero, and J. Belnap. 2014. Biological soil crusts across disturbance-recovery scenarios: effect of grazing regime on community dynamics. *Ecological Applications* 24: 1863–1877.

D'Antonio, C. M. and P. M. Vitousek. 1992. Biological invasions by exotic grasses, the grass/fire cycle, and global change. *Annual Review of Ecology & Systematics* 23: 63–87.

Desmond, M. 2004. Effects of grazing practices and fossorial rodents on a winter avian community in Chihuahua, Mexico. *Biological Conservation* 116: 235–242.

Diamond, J. M., C. A. Call, and N. Devoe. 2009. Effects of targeted cattle grazing on fire behavior of cheatgrass-dominated rangeland in the northern Great Basin, USA. *International Journal of Wildland Fire* 18: 944–950.

Diaz, S., S. Lavorel, S. McIntyre, V. Falczuk, F. Casanoves, D. G. Milchunas, C. Skarpe, G. Rusch, M. Sternberg, I. Noy-Meir, J. Landsberg, W. Zhang, H. Clark, and B. D. Campbell. 2007. Plant trait responses to grazing – a global synthesis. *Global Change Biology* 13: 313–341.

Dobkin, D. S., A. C. Rich, and W. H. Pyle. 1998. Habitat and avifaunal recovery from livestock grazing in a riparian meadow system of the northwestern Great Basin. *Conservation Biology* 12: 209–221.

Duniway, M. C., J. E. Herrick, and H. C. Monger. 2007. The high water-holding capacity of petrocalcic horizons. *Soil Science Society of America Journal* 71: 812–819.

Dziba, L. E., F. D. Provenza, J. J. Villalba, and S. B. Atwood. 2007. Supplemental energy and protein increase use of sagebrush by sheep. *Small Ruminant Research* 69: 203–207.

Elmore, W. and B. Kauffman. 1994. Riparian and watershed ecosystems: degradation and restoration. In M. Vavra, W. A. Laycock, and R. D. Pieper, editors, *Ecological Implications of Herbivory in the West*, 212–231. Denver, CO: Society for Range Management.

Frank, D. A. 2008. Ungulate and topographic control of nitrogen: phosphorus stoichiometry in a temperate grassland; soils, plants and mineralization rates. *Oikos* 117: 591–601.

Freilich, J. E., J. M. Emlen, J. J. Duda, D. C. Freeman, and P. J. Cafaro. 2003. Ecological effects of ranching: a six-point critique. *Bioscience* 53: 759–765.

Fuhlendorf, S. D., D. M. Engle, J. Kerby, and R. Hamilton. 2009. Pyric herbivory: rewilding landscapes through the recoupling of fire and grazing. *Conservation Biology* 23: 588–598.

Galloway, J. N., J. D. Aber, J. W. Erisman, S. P. Seitzinger, R. W. Howarth, E. B. Cowling, and B. J. Cosby. 2003. The nitrogen cascade. *Bioscience* 53: 341–356.

Gelbard, J. L. and J. Belnap. 2003. Roads as conduits for exotic plant invasions in a semiarid landscape. *Conservation Biology* 17: 420–432.

Gelbard, J. L. and S. Harrison. 2003. Roadless habitats as refuges for native grasslands: Interactions with soil, aspect, and grazing. *Ecological Applications* 13: 404–415.

Geremia, C., P. J. White, J. A. Hoeting, R. L. Wallen, F. G. R. Watson, D. Blanton, and N. T. Hobbs. 2014. Integrating population- and individual-level information in a movement model of Yellowstone bison. *Ecological Applications* 24: 346–362.

Grant, W. E., E. C. Birney, N. R. French, and D. M. Swift. 1982. Structure and productivity of grassland small mammal communities related to grazing-induced changes in vegetative cover. *Journal of Mammalogy* 63: 248–260.

Greenwood, K. L. and B. M. McKenzie. 2001. Grazing effects on soil physical properties and the consequences for pastures: a review. *Australian Journal of Experimental Agriculture* 41: 1231–1250.

Hanley, T. A. and J. L. Page. 1981. Differential effects of livestock use on habitat structure and rodent populations in Great Basin communities. *California Fish and Game* 68: 160–174.

Herrick, J. E., J. W. van Zee, K. M. Havstad, L. M. Burkett, and W. G. Whitford. 2009. *Monitoring Manual for Grassland, Shrubland and Savanna Ecosystems, Vol I*. Las Cruces, NM: USDA-ARS Jornada Experimental Range. Distributed by the University of Arizona Press, Tucson, AZ, USA.

Hobbs, N. T. 1996. Modification of ecosystems by ungulates. *Journal of Wildlife Management* 60: 695–713.

Hobbs, R. J. 2001. Synergisms among habitat fragmentation, livestock grazing, and biotic invasions in southwestern Australia. *Conservation Biology* 15: 1522–1528.

Hobbs, R. J. and L. F. Huenneke. 1992. Disturbance, diversity, and invasion: implications for conservation. *Conservation Biology* 6: 324–337.

Howery, L. E., J. E. Sprinkle, and J. E. Bowns. 2000. A summary of livestock grazing systems used on rangelands in the western United States and Canada. University of Arizona, Coop Extension AZ1184, Tucson http://hdl.handle.net/10150/144717

Hunt, L. P. 2001. Heterogeneous grazing causes local extinction of edible perennial shrubs: a matrix analysis. *Journal of Applied Ecology* 38: 238–252.

Huntly, N. 1991. Herbivores and the dynamics of communities and ecosystems. *Annual Review of Ecology and Systematics* 22: 477–503.

IPCC. 2018. Global Warming of 1.5°C. An IPCC Special Report on the impacts of global warming of 1.5°C above pre-industrial levels and related global greenhouse gas emission pathways, in the context of strengthening the global response to the threat of climate change, sustainable development, and efforts to eradicate poverty (V. Masson-Delmotte, P. Zhai, H. O. Pörtner, D. Roberts, J. Skea, P. R. Shukla, A. Pirani, W. Moufouma-Okia, C. Péan, R. Pidcock, S. Connors, J. B. R. Matthews, Y. Chen, X. Zhou, M. I. Gomis, E. Lonnoy, T. Maycock, M. Tignor, T. Waterfield, editors).

Kauffman, J. B. and W. C. Krueger. 1984. Livestock impacts on riparian ecosystems and streamside management implications – a review. *Journal of Range Management* 37: 430–438.

Kikoti, I. A. and C. Mligo. 2015. Impacts of livestock grazing on plant species composition in montane forests on the northern slope of Mount Kilimanjaro, Tanzania. *International Journal of Biodiversity Science Ecosystem Services & Management* 11: 114–127.

Kimball, S., J. R. Gremer, G. A. Barron-Gafford, A. L. Angert, T. E. Huxman, and D. L. Venable. 2014. High water-use efficiency and growth contribute to success of non-native Erodium cicutarium in a Sonoran Desert winter annual community. *Conservation Physiology* 2 , cou006.

Koerner, S. E., M. D. Smith, D. E. Burkepile, N. P. Hanan, M. L. Avolio, S. L. Collins, A. K. Knapp, N. P. Lemoine, E. J. Forrestel, S. Eby, D. I. Thompson, G. A. Aguado-Santacruz, J. P. Anderson, T. M. Anderson, A. Angassa, S. Bagchi, E. S. Bakker, G. Bastin, L. E. Baur, K. H. Beard, E. A. Beever, P. J. Bohlen, E. H. Boughton, D. Canestro, A. Cesa, E. Chaneton, J. Cheng, C. M. D'Antonio, C. Deleglise, F. Dembélé, J. Dorrough, D. J. Eldridge, B. Fernandez-Going, S. Fernández-Lugo, L. H. Fraser, B. Freedman, G. García-Salgado, J. R. Goheen, L. Guo, S. Husheer, M. Karembé, J. M. H. Knops, T. Kraaij, A. Kulmatiski, M.-M. Kytöviita, F. Lezama, G. Loucougaray, A. Loydi, D. G. Milchunas, S. J. Milton, J. W. Morgan, C. Moxham, K. C. Nehring, H. Olff, T. M. Palmer, S. Rebollo, C. Riginos, A. C. Risch, M. Rueda, M. Sankaran, T. Sasaki, K. A. Schoenecker, N. L. Schultz, M. Schütz, A. Schwabe, F. Siebert, C. Smit, K. A. Stahlheber, C. Storm, D. J. Strong, J. Su, Y. V. Tiruvaimozhi, C. Tyler, J. Val, M. L. Vandegehuchte, K. E. Veblen, L. T. Vermeire, D. Ward, J. Wu, T. P. Young, Q. Yu, and T. J. Zelikova. 2018. Change in dominance determines herbivore effects on plant biodiversity. *Nature Ecology & Evolution* 2: 1925–1932.

Lauenroth, W. K., D. G. Milchunas, J. L. Dodd, R. H. Hart, R. K. Heitschmidt, and L. R. Rittenhouse. 1994. Effects of grazing on ecosystems of the Great Plains. In M Vavra, W. A. Laycock, R. D. Pieper, editors, *Ecological Implications of Livestock Herbivory in the West*, 69–100. Denver, CO: Society for Range Management.

Leu, M., Hanser, S. E. and Knick, S. T. 2008. The human footprint in the West: a large-scale analysis of anthropogenic impacts. *Ecological Applications* 18: 1119–1139.

Levin, P. S., J. Ellis, R. Petrik, and M. E. Hay. 2002. Indirect effects of feral horses on estuarine communities. *Conservation Biology* 16: 1364–1371.

Lind, E. M., E. Borer, E. Seabloom, P. Adler, J. D. Bakker, D. M. Blumenthal, M. Crawley, K. Davies, J. Firn, D. S. Gruner, W. Stanley Harpole, Y. Hautier, H. Hillebrand, J. Knops, B. Melbourne, B. Mortensen, A. C. Risch, M. Schuetz, C. Stevens, and P. D. Wragg. 2013. Life-history constraints in grassland plant species: a growth-defence trade-off is the norm. *Ecology Letters* 16: 513–521.

Liu, Y. S., Q. M. Pan, H. D. Liu, Y. F. Bai, M. Simmons, K. Dittert, and X. G. Han. 2011. Plant responses following grazing removal at different stocking rates in an Inner Mongolia grassland ecosystem. *Plant and Soil* 340: 199–213.

Lovich, J. E. and D. Bainbridge. 1999. Anthropogenic degradation of the southern California desert ecosystem and prospects for natural recovery and restoration. *Environmental Management* 24: 309–326.

Loydi, A. and S. M. Zalba. 2009. Feral horses dung piles as potential invasion windows for alien plant species in natural grasslands. *Plant Ecology* 201: 471–480.

Mack, R. N. and J. N. Thompson. 1982. Evolution in steppe with few large, hoofed mammals. *American Naturalist* 119: 757–773.

McGovern, S. T., C. D. Evans, P. Dennis, C. A. Walmsley, A. Turner, and M. A. McDonald. 2014. Increased inorganic nitrogen leaching from a mountain grassland ecosystem following grazing removal: a hangover of past intensive land-use? *Biogeochemistry* 119: 125–138.

McPherson, G. R. 1995. The role of fire in the desert grasslands. In M. P. McClaran and T. R. Van Devender, editors, *The Desert Grassland*, 130–151. Tucson, AZ: University of Arizona Press.

Merkle, J. W., K. L. Monteith, E. O. Aikens, M. M. Hayes, K. R. Hersey, A. D. MIddelton, B. A. Oates, H. Sawyer, B. M. Scurlock, and M. J. Kauffman. 2016. Large herbivores surf waves of green-up during spring. *Proceedings of the Royal Society B: Biological Sciences* 283: https://doi.org/10.1098/rspb.2016.0456

Milchunas, D. G. and W. K. Lauenroth. 1993. Quantitative effects of grazing on vegetation and soils over a global range of environments. *Ecological Monographs* 63: 327–366.

Milchunas, D. G. and I. Noy-Meir. 2002. Grazing refuges, external avoidance of herbivory and plant diversity. *Oikos* 99: 113–130.

Milchunas, D. G., O. E. Sala, and W. K. Lauenroth. 1988. A generalized model of the effects of grazing by large herbivores on grassland community structure. *American Naturalist* 132: 87–106.

Miller, R. F., J. C. Chambers, D. A. Pyke, F. B. Pierson, and C. J. Williams. 2013. A review of fire effects on vegetation and soils in the Great Basin Region: response and ecological site characteristics. U.S. Department of Agriculture, Forest Service, Rocky Mountain Research Station, General Technical Report RMRS-GTR-308, Fort Collins, CO. 126 pages.

Miller, R. F. and E. K. Heyerdahl. 2008. Fine-scale variation of historical fire regimes in sagebrush-steppe and juniper woodland: an example from California, USA. *International Journal of Wildland Fire* 17: 245–254.

Miller, R. F. and P. E. Wigand. 1994. Holocene changes in semiarid pinyon-juniper woodlands. *Bioscience* 44: 465–474.

Monroe, A. P., C. L. Aldridge, T. J. Assal, K. E. Veblen, D. A. Pyke, and M. L. Casazza. 2017. Patterns in Greater Sage-grouse population dynamics correspond with public grazing records at broad scales. *Ecological Applications* 27: 1096–1107.

Moser, B. W. 1997. *The Effects of Elk and Cattle Grazing on the Vegetation, Birds, and Small Mammals of the Bridge Creek Wildlife Area, Oregon.* Pullman, WA: Washington State University.

Mote, P. W. and K. T. Redmond. 2011. Western climate change. In E. A. Beever and J. Belant, editors, *Ecological Consequences of Climate Change: Mechanisms, Conservation, and Management.* Boca Raton, FL: CRC Press (Taylor and Francis Group). 314pp.

Munson, S. M., R. H. Webb, D. C. Housman, K. E. Veblen, K. E. Nussear, E. A. Beever, K. B. Hartney, M. N. Miriti, S. L. Phillips, R. E. Fulton, and N. G. Tallent. 2015. Long-term plant responses to climate are moderated by biophysical attributes in a North American desert. *Journal of Ecology* 103: 657–668.

Nowak, R. S., D. S. Ellsworth, and S. D. Smith. 2004. Functional responses of plants to elevated atmospheric CO_2 – do photosynthetic and productivity data from FACE experiments support early predictions? *New Phytologist* 162: 253–280.

Oesterheld, M. and S. J. McNaughton. 1991. Effect of stress and time for recovery on the amount of compensatory growth after grazing. *Oecologia* 85: 305–313.

Oldemeyer, J. L. 1994. Livestock grazing and the desert tortoise in the Mojave Desert. *U.S. Fish and Wildlife Service Fish and Wildlife Research* 13: 94–103.

Paige, K. N. 1999. Regrowth following ungulate herbivory in *Ipomopsis aggregata*: geographic evidence for overcompensation. *Oecologia* 118: 316–323.

Parisien, M.-A., S. Snetsinger, J. A. Greenberg, C. R. Nelson, T. Schoennagel, S. Z. Dobrowski, and M. A. Moritz. 2012. Spatial variability in wildfire probability across the western United States. *International Journal of Wildland Fire* 21: 313–327.

Perevolotsky, A. and N. G. Seligman. 1998. Role of grazing in Mediterranean rangeland ecosystems: Inversion of a paradigm. *Bioscience* 48: 1007–1017.

Peters, D. P. C. and R. P. Gibbens. 2006. Plant communities in the Jornada Basin: the dynamic landscape. In K. Havstad, L. F. Huenneke, and W. H. Schlesinger, editors, *Structure and Function of Chihuahuan Desert Ecosystem: the Jornada Basin Long-term Ecological Research Site,* 211–231. New York: Oxford University Press.

Peters, D. P. C., R. A. Pielke, B. T. Bestelmeyer, C. D. Allen, S. Munson-McGee, and K. M. Havstad. 2004. Cross-scale interactions, nonlinearities, and forecasting catastrophic events. *Proceedings of the National Academy of Sciences of the United States of America* 101: 15130–15135.

Peterson, G., C. R. Allen, and C. S. Holling. 1998. Ecological resilience, biodiversity, and scale. *Ecosystems* 1: 6–18.

Polley, H. W., D. D. Briske, J. A. Morgan, K. Wolter, D. W. Bailey, and J. R. Brown. 2013. Climate change and North American rangelands: trends, projections, and implications. *Rangeland Ecology & Management* 66: 493–511.

Popotnik, G. J. and W. M. Giuliano. 2000. Response of birds to grazing of riparian zones. *Journal of Wildlife Management* 64: 976–982.

Porensky, L. M. and K. E. Veblen. 2015. Generation of ecosystem hotspots using short-term cattle corrals in an African savanna. *Rangeland Ecology & Management* 68: 131–141.

Powers, L. C. and H. A. Glimp. 1996. Impacts of livestock on shorebirds: a review and application to shorebirds of the western Great Basin. *International Wader Studies* 9: 55–63.

Pyke, D. A. 2011. Restoring and rehabilitating sagebrush habitats. In S. T. Knick and J. W. Connelly, editors, *Studies in Avian Biology, 38, Ecology and Conservation of a Landscape Species and its Habitats*, 531–548. Berkeley, CA: University of California Press.

Pyke, D. A., M. L. Brooks, and C. D'Antonio. 2010. Fire as a restoration tool: a decision framework for predicting the control or enhancement of plants using fire. *Restoration Ecology* 18: 274–284.

Pyke, D. A., J. C. Chambers, J. L. Beck, M. L. Brooks, and B. A. Mealor. 2016. Land uses, fire and invasion: exotic annual Bromus and human dimensions. In M. J. Germino, J. C. Chambers, and C. S. Brown, editors, *Exotic Brome-grasses in Arid and Semiarid Ecosystems of the Western US: Causes, Consequences, and Management Implications*. Berlin, Germany: Springer.

Reisner, M. D., J. B. Grace, D. A. Pyke, and P. S. Doescher. 2013. Conditions favouring *Bromus tectorum* dominance of endangered sagebrush steppe ecosystems. *Journal of Applied Ecology* 50: 1039–1049.

Schulz, T. T. and W. C. Leininger. 1991. Nongame wildlife communities in grazed and ungrazed montane riparian sites. *Great Basin Naturalist* 51: 286–292.

Seefeldt, S. S. and A. B. Leytem. 2011. Sheep bedding in the Centennial Mountains of Montana and Idaho: effects on vegetation. *Western North American Naturalist* 71: 361–373.

Smith, S. D., T. N. Charlet, S. F. Zitzer, S. R. Abella, C. H. Vanier, and T. E. Huxman. 2014. Long-term response of a Mojave Desert winter annual plant community to a whole-ecosystem atmospheric CO2 manipulation (FACE). *Global Change Biology* 20: 879–892.

Strand, E. K., K. L. Launchbaugh, R. F. Limb, and L. A. Torell. 2014. Livestock grazing effects on fuel loads for wildland fire in sagebrush dominated ecosystems. *Journal of Rangeland Applications* 1: 35–37.

Strauss, S. Y. and A. A. Agrawal. 1999. The ecology and evolution of plant tolerance to herbivory. *Trends in Ecology & Evolution* 14: 179–185.

Tagestad, J., M. Brooks, V. Cullinan, J. Downs, and R. McKinley. 2016. Precipitation regime classification for the Mojave Desert: implications for fire occurrence. *Journal of Arid Environments* 124: 388–397.

Tewksbury, J. J., A. E. Black, N. Nur, V. A. Saab, B. D. Logan, and D. S. Dobkin. 2002. Effects of anrthropogenic fragmentation and livestock grazing on western riparian bird communities. *Studies in Avian Biology* 25: 258–202.

Trimble, S. W. and A. C. Mendel. 1995. The cow as a geomorphic agent – a critical review. *Geomorphology* 13: 233–253.

Van Auken, O. W. 2009. Causes and consequences of woody plant encroachment into western North American grasslands. *Journal of Environmental Management* 90: 2931–2942.

Van Devender, T. R. 1995. Desert grassland history: changing climates, evolution, biogeography, and community dynamics. In M. P. McClaran and T. R. Van Devender, editors, *The Desert Grassland*, 68–99. Tucson, AZ: University of Arizona Press.

Van Haveren, B. P. 1983. Soil bulk density as influenced by grazing intensity and soil type on a shortgrass prairie site. *Journal of Range Management* 36: 586–588.

Veblen, K. E. 2012. Savanna glade hotspots: plant community development and synergy with large herbivores. *Journal of Arid Environments* 78: 119–127.

Veblen, K. E., K. C. Nehring, C. M. McGlone, and M. E. Ritchie. 2015. Contrasting effects of different mammalian herbivores on sagebrush plant communities. *PLOS ONE* 10: e0118016.

Veblen, K. E., L. M. Porensky, C. Riginos, and T. P. Young. 2016. Are cattle surrogate wildlife? Savanna plant community composition explained by total herbivory more than herbivore type. *Ecological Applications* 26: 1610–1623.

Walker, B. H., N. Abel, D. M. Stafford Smith, and J. L. Langridge. 2002. A framework for the determinants of degradation in arid ecosystems. In J. F. Reynolds and D. M. Stafford Smith, editors, *Global Desertification: Do Humans Cause Deserts?*, 75–94. Berlin, Germany: Dahlem University Press.

Warren, S. D. and D. J. Eldridge. 2001. Biological soil crusts and livestock in arid ecosystems: are they compatible? In J. Belnap and O. L. Lange, editors, *Biological Soil Crusts: Structure, Function, and Management*, 403–415. Berlin: Springer.

Webb, R. H. 2002. Recovery of severely compacted soils in the Mojave Desert, California, USA. *Arid Land Research and Management* 16: 291–305.

Weltz, M., M. K. Wood, and E. E. Parker. 1989. Flash grazing and trampling – effects on infiltration rates and sediment yield on a selected New Mexico range site. *Journal of Arid Environments* 16: 95–100.

Yan, R., X. Xin, Y. Yan, X. Wang, B. Zhang, G. Yang, S. Liu, Y. Deng, and L. Li. 2015. Impacts of differing grazing rates on canopy structure and species composition in hulunber meadow steppe. *Rangeland Ecology & Management* 68: 54–64.

chapter five

Response of Tick-borne Disease to Fire and Timber Harvesting

Mechanisms and Case Studies across Scales

Janet E. Foley and Benjamin T. Plourde

Introduction

Anthropogenic disturbances of all types and intensities have the potential to modify the risk of infectious disease (Gottdenker et al. 2014). On one extreme, ecosystems are intentionally repurposed for human development. The construction of dams and roads can have extensive impacts on plant and animal communities while exploitation of natural resources (e.g. for food, timber, minerals, or other resources) varies in intensity. Climate and fire are distinctive, far-reaching drivers because they are often natural and often uncontrolled, yet increasingly influenced by human activity. In addition to the intensity of the disturbance, a second "axis" of human influence on natural lands is spatial scale, with climate having the greatest scope of all of these, affecting the entire planet. The spatial extent of other disturbances depends on not only the size of the disturbance, but also downstream influences due to connectivity with surrounding areas. Components of communities in affected and downstream patches may become rare or locally extirpated, symbioses may be decoupled, some organisms may be released from regulation and expand in number and range, and some may change their natural patterns to occupy new niches in the absence of resources, predators, parasites, and competitors.

A broad view of mechanisms by which ecosystem disturbances affect disease has advantages for managers who would like to predict or mitigate negative health impacts. Such a synthesis has been elusive because scientists typically work in areas of specialty, there is a lack of data on many systems, and the scientific community tends to focus on details and specific outcomes of disturbance but less on common processes. Here, we work through one example (maintenance of tick-borne disease) of a multiple-trophic system and its response to two widespread types of disturbance. We first propose hypotheses and mechanisms by which disturbance can influence tick-borne disease. Second, we summarize the current research on the response of tick-borne diseases to timber harvesting and fire. The third step is to consider missing steps along causal pathways that have not yet been addressed in the literature. Finally, we organize data gaps that limit progress towards a broad synthesis of disturbance and tick-borne disease. Throughout, we assess the extents to which temporal and spatial scales, ecological contingency, and specific biological details of the organisms involved impede our ability to generalize about outcomes following disturbance.

Tick-borne Disease

After mosquitoes, ticks are the second-most-important vector of human and animal diseases worldwide (de Castro 1997). Ticks transmit agents of Lyme disease, anaplasmosis, heartwater, Rocky Mountain spotted fever, tick-borne encephalitis, ehrlichiosis, relapsing fever, and other serious diseases (Brown et al. 2005). The geographical ranges of some ticks and tick-borne diseases are expanding, particularly in cold areas, purportedly due to global climate change (Randolph and Rogers 2000; Ogden et al. 2006), expanding global travel and trade (Burridge et al. 2000; Owen et al. 2006), and anthropogenic habitat change (Allan et al. 2003).

Tick-borne disease is an inherently multi-trophic system depending on ticks, their hosts, microbes, and their biotic surroundings. Thus, environmental changes that affect any component of the system can be expected to change disease risk. However, the possible effects of these changes are likely as diverse as the ecosystems and ticks that maintain pathogens. We can better predict the responses of tick-borne disease to ecosystem disturbance by considering how different disturbance types may affect each of the components of the tick-borne disease system.

Hard ticks (members of the family Ixodidae, which are the more prominent and better studied vectors of disease) have relatively long life spans, are obligately hematophagous parasites, and feed only once in each of the three host-feeding stages (larva, nymph, adult). The time in between feeding (typically off-host) is spent digesting the blood meal, molting, and hardening off prior to host seeking. Such inter-host intervals can span months to a year, meaning that even pathogens that have already colonized and killed or induced herd immunity in a large proportion of a host population (and thus might otherwise be facing local extinction) may persist in a reservoir of off-host ticks.

The physical environments of off-host ticks are important determinants of tick geographical ranges, abundance, and survival. While off hosts, "questing" species of ticks reside in litter, foraying onto vegetation where they await passing hosts. Especially while questing, they are vulnerable to desiccation and while in litter they are vulnerable to fungal infection. Increases in ambient temperature and extremes of humidity are associated with reduced tick survival and tick numbers (Bertrand and Wilson 1996). In contrast, some nidicolous (nest-dwelling) tick species live within animal burrows where microenvironments are less variable. These species are then dependent on host use of the burrows. Tick species also vary in host range from a single species to dozens (McCoy et al. 2013). Given that disturbances often change vertebrate community compositions, the survival of ticks may equally depend on the availability of suitable hosts.

Ecosystem Disturbance and Tick-borne Disease

We focus our analysis on fire and timber harvesting because these occur repeatedly, have important impacts on tick-borne disease, and are relatively under-studied in this context. Such disturbances can increase or decrease local infectious-disease risk. Hypotheses for both outcomes are represented in Figure 5.1. In Figure 5.1a, a scenario dubbed "the cruel forest," the intact, heterogeneous ecosystem provides all of the ecological needs of pathogens and vectors, thus promoting disease. Habitat disturbance could disrupt pathogen maintenance cycles and reduce disease risk (Figure 5.1a). "The kind forest" is also heterogeneous and supports a diversity of hosts that keeps pathogen prevalence low. Disturbance may homogenize the landscape while concomitantly selecting for hosts central to pathogen maintenance and increasing disease risk (Figure 5.1b). On the contrary, an undisturbed system may comprise more homogeneous habitat and have low biodiversity ("the simple forest," Figure 5.1c). Moderate disturbance can serve to increase habitat heterogeneity, thus increasing diversity and decreasing pathogen abundance (Figure 5.1c). Across a wide variety of systems, parasite abundance is negatively related to host biodiversity (Civitello et al. 2015). In the face of increased diversity, infectious disease risk may be lowered by one of several potential mechanisms such as diversion of blood meals to non-competent hosts, reducing competent host numbers through predation or competition, altering host behaviors, etc. Finally, if natural ecosystems tend to be more isolated from people and domestic animals ("the hidden forest"), then disturbance may remove buffers (e.g. through road-building) and increase disease risk (Figure 5.1b). Throughout our examination of tick-borne disease response to disturbance, we find support for all four hypotheses.

In addition, we outline causal pathways for how disturbance causes changes in tick-borne disease risk (Figure 5.2). Depending on scale, a disturbance will often produce edges and modify the complexity of the landscape, both of which are especially important to vector-borne disease ecology (Allan et al. 2003, Molyneux 2003). At the coarsest scale, a patchwork landscape of disturbed, transitional, and developed areas will regulate populations of large mammals and determine the risk of disease. By considering how forest management and fire impact each step in the causal pathway, we hope to be able to predict outcomes and eventually offer insights into links in the causal chain that could be targets of management to reduce risk.

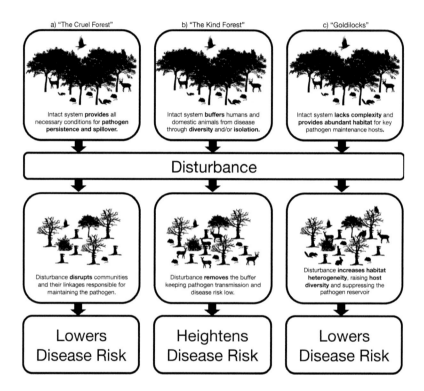

Figure 5.1 Hypotheses to explains changes in disease risk following disturbance. (a) Following the cruel forest hypothesis, disturbance removes one or more components responsible for the maintenance of a pathogen. Infectious disease risk wanes with disturbance. (b) The kind forest harbors pathogens, but keeps risk low by supporting a diversity of hosts that dilutes pathogen prevalence. Disturbance reduces this ecosystem service, increasing disease risk by creating beneficial habitat for key reservoir hosts. (c) Alternatively, in a forest that is "just right" (Goldilocks), disturbance can increase local diversity by increasing habitat heterogeneity, creating niches for otherwise limited species. If the conditions of the dilution effect are met, the disturbed habitat presents a lower disease risk. An isolated forest serves to buffer the risk of infectious disease through its inaccessibility. Disturbance, through road-building or other infrastructure, increases overlap between reservoir and target, heightening risk.

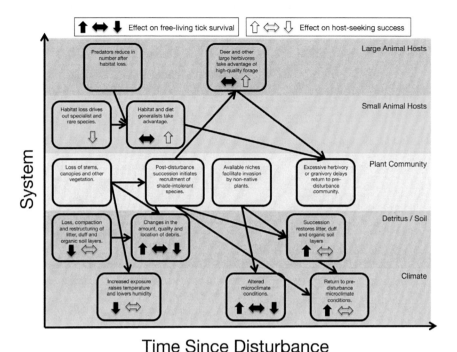

Time Since Disturbance

Figure 5.2 Causal pathways determining tick response following restructuring disturbances. The response of ticks to disturbance is ultimately dependent on free-living tick survival and their host-seeking success. Plant and animal communities should respond to disturbance, each directly or indirectly modifying the tick response. Interactions among the disturbance, climate, and communities will dictate the short- and long-term responses outlined here.

Fire and Timber Harvesting

Fire affects both forests and grasslands, imposing variable degrees of habitat change, from thinning understory and debris to complete eradication of communities. Natural and some intentional fires used by earlier human societies are thought to have been important for the persistence of some types of prairie and redwood (*Sequoia sempervirens*)-dominated forests (Hulbert 1986; Williams 2003; Lorimer et al. 2009). Increasingly, fire is influenced by humans through arson, prescribed burning, and fire suppression, which can facilitate high-intensity wildfires (Roberts et al. 2015). High-intensity fire differs quite considerably from low-intensity fire in North American forests in terms of succession with fire-dependent species, downstream biodiversity, and ecosystem functioning (Fontaine and Kennedy 2012; DellaSala and Hanson 2015). Indirect anthropogenic influences on fire include climate change (Abatzoglou and Williams 2016;

Schoennagel et al. 2017) and human activities that alter fire probabilities (e.g. suppression and road building) (Balch et al. 2017). There is an increasing literature on fire and its role in structuring communities, although to date most studies still consider relatively small spatial scales or short time intervals (Fontaine and Kennedy 2012).

The influence of fire on tick-borne disease has received only a small amount of attention in the scientific literature (Scasta 2015), consisting primarily of studies that report the effects of small and low-intensity prescribed burns on ticks directly. Because data from long-term studies are sparse, deriving general predictions about fire and tick-borne disease is difficult. Further, community assembly after fire may not be the same across different ecosystems, and demonstrates considerable idiosyncrasies. Size, intensity, edginess, and proximity of the burned area to a source of colonizers can all influence the successional dynamics and the prevalence of ticks, hosts, and tick-borne pathogens.

Fire is the combustion of flammable material that can convert organic matter into ash (carbon, phosphorus, and potassium debris, among others), heat energy, carbon dioxide, and water. Burning often releases toxic compounds into the atmosphere. Organisms may die during a fire due to heat, lack of oxygen, and/or toxicosis. All three can kill vertebrates but likely only heat kills ticks. If a burned area is sufficiently large and the fire sufficiently intense, ticks and hosts may be transiently eradicated, although in many cases, ticks and their hosts survive low-intensity fires. For example, prescribed burning of lone star tick (*Amblyomma americanum*) habitat in Oklahoma achieved temperatures of 110°C, sufficient to kill ticks in leaf litter (Hoch et al. 1972). Temperatures deeper in the duff were much lower and many ticks were still present after the burn. In fact, a year later, nymph abundance exceeded pre-burn levels. In contrast, lone star tick (*Amblyomma americanum*) numbers were reduced in years when tall-grass prairie burned and one-year burn intervals were associated with fewer ticks than longer times between burns (Cully 1999). Other studies that document reduction in tick numbers shortly after fire include lone star ticks in Georgia piedmont (Davidson et al. 1994), deer ticks (*Ixodes scapularis*) in northern Florida (Rogers 1955), and winter ticks (*Dermacentor albipictus*) in Alberta (Drew et al. 1985), among others. Although not apparent from the publications, one mediating difference could be the intensity of the fires (Table 5.1). For example, chaparral in California is heavily influenced by fire and many community members are adapted to fire cycles. Thus, while one year after prescribed burns in chaparral, ticks held in below-ground packets survived the relatively low-intensity burn regime and rodent numbers were reduced with fewer large rodents (specifically woodrats, *Neotoma fuscipes*) present, tick abundance was either the same or higher (Padgett et al. 2009).

Table 5.1 Summary of current knowledge, impact of scale, and open questions concerning tick response to fire and timber harvesting

Disturbance	Patterns	Scale	Open questions
Fire	Tick abundance is lower in the months following prescribed burns. Tick recovery after burns is driven, at least in part, by high quality forage attracting large herbivores to disturbed sites. Small mammal community is more abundant following fire.	Prescribed burning studies cover areas from 0.2 to 2500 hectares. Most studies consider burns <100 ha. Wildfires of any size are largely unstudied. Most studies sample ticks at a single time point shortly after burning. The trajectory back to the pre-disturbance state is unknown.	Do findings from prescribed burn studies translate to wildfires of higher intensity and larger size? What is the balance between remnant tick survival and host-carried ticks in restoring the tick community? How does connectivity to unburned patches influence the response? How does the timing of fire (seasonally and time of day) affect tick survival?
Timber Harvesting	Canopy openings lead to recruitment of high-quality forage, leading to increased usage by cervids. Forest fragmentation across landscapes impacts tick-borne disease risk and incidence, but mechanisms are unknown. Small mammal generalists take advantage of abundant resources on logged patches.	Studies to date consider tick and pathogen response to patterns of disturbance across landscapes, often at the municipality, county, and country levels. Degree of fragmentation, edginess and land cover classes are common explanatory variables. Short-term and fine-scale response of ticks to logging is unknown, but potentially predictable given many related studies.	How does tree felling and extraction affect forest floor microclimates? What is the relationship between logging intensity and the response of ticks? Can uneven-aged management increase habitat heterogeneity, increase host diversity, and lower parasite abundance?

Fire and logging can both initiate similar patterns of community succession and likewise similarly influence disease. Forests are in constant cycles of change and harbor many tick-borne pathogens, including those infecting humans. Anthropogenic disturbances of forest ecosystems range from food gathering and hunting to deforestation. Thirty percent of the world's forests are actively managed for production and another quarter are classified as mixed-use (FAO 2015). Timber harvesting varies from single tree selection and other uneven management schemes to even-aged clear-cuts. Deforestation removes trees and is followed by conversion to another land use (e.g. urbanization or agriculture). Deforestation to create agricultural land by slash-and-burn techniques is a major source of habitat change worldwide. Approximately 6.6 million hectares of the world's natural forests are lost annually (FAO 2015), although this may be a conservative estimate, given differences in forest definition and reporting practices.

Although there are many studies on disturbance and its impact on vector-borne disease, most emphasize insects (Gottdenker et al. 2014). A general understanding of tick and pathogen response to ecosystem disturbances is impeded by a paucity of studies and our inability to directly translate impacts on diseases vectored by insects to tick-borne disease because ticks have such distinct life histories.

Timber harvesting differs from fire in its immediate impact on tick microenvironments (Figure 5.2). While fire can eliminate the litter and duff layers, logging operations are more likely to a) increase litter volume with slash, and b) compact soil with machinery. However, as the ecosystem undergoes post-fire or post-logging succession, common themes emerge. Emergence of tick-borne pathogens requires assembly of a community that can support hosts of juvenile ticks (commonly rodents, insectivores, or lizards), hosts of adult ticks (larger mammals such as cervids and carnivores), and free-living ticks (Figure 5.2). If the nearest source of colonizing organisms is disturbed, the new ecosystem that emerges may be very different from the one that burned (Figure 5.1).

Early-stage impacts on tick-borne disease are heavily mediated by the disturbance's impacts on the plant community (Figure 5.2). Many questing tick species ascend vegetation, particularly as adults, to seek hosts for a blood meal. At least temporarily, any adult ticks that survive a fire will not have vegetation to assist in their host-seeking. Larval and nymphal ticks utilize vegetation to a lesser degree for host-seeking. Clearly, loss of many or all components of plant communities will have immediate and devastating impacts on herbivorous vertebrates and cascading impacts on predators. Subsequent to fire or clear-cutting, a common trajectory in plant communities is rapid growth of shade-intolerant species, followed by the gradual return of the late-successional community. Although native species adapted to post-fire emergence may reappear, fire can facilitate the invasion of exotic species taking advantage of newly available niches

(Harrod and Reichard 2000). In parallel with the regrowth of the plant community, particular animals recolonize sites early, such as deer mice (*Peromyscus maniculatus*) and deer (*Odocoileus* spp.) in North America. In fact, the abundance of fresh plant growth can be highly favorable for herbivore population growth. Burned sites in the Sierra Nevada had higher habitat heterogeneity and increased small mammal community evenness (Roberts et al. 2015), suggesting that fire may also act to increase diversity. These changes differed with differing fire intensity. Similarly, there was some evidence for higher small mammal richness and diversity along 12- and 19-year old logging extraction routes (Malcolm and Ray 2000). Although that study's focus was not infectious disease, these changes could lead to the dilution of a pathogen (Figure 5.1c). The suppression of fire may likewise alter natural cycles and influence disease cycles.

Although fire can cause an immediate reduction in ticks and tick-borne pathogens, the dynamic processes that follow fire could promote *increased* tick density and tick-borne pathogen prevalence (Figure 5.2). Examples of this are increased deer recruitment associated with increased ticks after prescribed burning in Oklahoma (Hoch et al. 1972), recovery of ticks as early as five months post-burn via deer and white-footed mice (*P. leucopus*) (Stafford III et al. 1998), and increased deer and lone star tick numbers one year after burns in the Ozark Mountains (Allan 2009). Prescribed burning in Long Island, New York halved tick density, but effected no change in the risk of Lyme disease as measured by the density of infected nymphs (Mather et al. 1993).

Ecosystem disturbance could dramatically influence tick-borne pathogen prevalence through changes in community composition. Ticks vary as to whether they are host specialists or generalists (McCoy et al. 2013). Some important vectors of zoonotic disease are generalists, feeding on small mammals, reptiles, or birds in juvenile stages and larger vertebrates including humans as nymphs and adults (Brown et al. 2005). The relative abundances of reptiles and mammals can influence the ability of a community to maintain tick-borne pathogens. An excellent example is the maintenance of *Borrelia burgdorferi* in the western United States. The vector, *Ixodes pacificus*, commonly infests fence (*Sceloporus* spp.) and alligator (*Elgaria* spp.) lizards, but reptiles are not competent hosts of the pathogen and in fact protein components in their blood may kill the pathogen and reduce overall community prevalence (Lane and Quistad 1998). Changes in the composition of these communities after disturbance have not been studied simultaneously, but experimental manipulations and observational work show that specific community composition is very important in determining tick-borne disease risk (Salkeld and Lane 2010; Swei et al. 2011b).

The influence of disturbance on tick-borne disease will manifest differently across spatial scales. Most hard ticks cannot move even a few meters unless on hosts, so their distribution is constrained by the movements of hosts. The only two plausible sources of tick recolonization on disturbed lands are large mammals and birds. Of importance, both large mammals and birds will likely only recolonize a site once an appropriate plant community is established. Large mammals may bring fed adult ticks that will detach and lay eggs. A microenvironment with sufficient humidity is required for survival of eggs and larvae, which will typically emerge after months or a few seasons to seek a larval blood meal from small mammals, reptiles, or birds. A number of birds have been shown to be competent hosts for ticks. Although birds typically have low tick loads (number of ticks per birds), infestation is significant if birds contribute to long-distance tick dispersal. Some ground-foraging migratory birds shown to host high intensities of ticks or even be infected with vector-borne pathogens include American robins (*Turdus migratorius*), dark-eyed juncos (*Junco hyemalis*), and golden-crowned sparrows (*Zonotrichia atricapilla*), among many others (Dingler et al. 2014). Long-distance spread of *Borrelia* spp. via *Ixodes uriae* has been reported on seabirds (Olsen et al. 1995).

Case Studies of Tick-borne Disease Response to Disturbance

Congolese Ticks

Africa is sometimes described as "the Continent of Fire," because natural wildfire is an important component of community dynamics in many areas. In addition, anthropogenic fires have been used in slash-and-burn agriculture and to drive game distributions in many areas. In the Democratic Republic of Congo (DRC), profound social unrest followed by a decrease in conflict led to the return of over a million people in the early 2000s. Although the DRC is a large country with immense natural resources, lack of infrastructure and food prompted acceleration of slash-and-burn practices in rural areas to meet demands for food and wood (Iloweka 2004). We recently studied the relationship between these practices and tick-borne disease in the rural communities of Kikongo and Kimbuma, 200 km east of Kinshasa in the Bandundu province. Although intact Western Congolian forest-savanna mosaic can be found within western DRC, burning and human incursion are converting forest to savanna, increasing the overall area that is ecotone, i.e. grassland abutting forest with the Congolian lowland rainforest, and reducing wildlife diversity and abundance. Our study area is characterized by valleys containing undisturbed rainforest, plateaus with disturbed grasslands, and gardens in clearings. Parts of the savannas have been further cleared by fire to create gardens. The pace of conversion to savanna is increasing rapidly (Iloweka 2004). Fire continues to be used extensively across the DRC to create pasture and to drive game out during hunting (MODIS Rapid Response Team 2010).

The DRC has numerous vector-borne diseases, including malaria, river blindness, sleeping sickness, and numerous viral pathogens (Rimoin and Hotez 2013). There is a strong bias of African tick collections towards parasites of cattle (Cumming 2000), and less is known about ticks affecting humans and other animals (Burt et al. 1996). Medically important ticks found in the DRC include *Haemaphysalis leachi*, a vector of *Babesia canis* (Jefferies et al. 2009), *Rickettsia conorii* (Hoogstraal 1971), and *Rhipicephalus muhsamae* that also transmits rickettsiae (Parola et al. 2005).

We had an opportunity to prospectively examine the distribution of ticks in areas where clearing had been done for agriculture in savannas that were subjected to fire to drive game, including rodents. Ticks were collected during the dry season in July and August 2008 in unburned forest control sites, burned fields, and manioc gardens in forest clearings. Sixty 10-m transects in each habitat were sampled from 07:00–10:00, and again from 15:00–18:00. A worker would walk along the transect dragging a 1 m^2 white cloth, wait one minute, and then walk along the transect with the cloth a second time collecting visible, questing ticks by hand from the cloth, ground, or vegetation into vials (one for each transect) containing 70 percent ethanol. Villagers were hired so that all three habitats could be surveyed at the same time on the same day. Ticks were identified with respect to sex, stage, and species using a published key (Matthysse and Colbo 1987). All of the ticks collected were adult *Rhipicephalus cliffordi* ($N=179$) and *H. leachi* (210). Nymphs and larvae may have been overlooked because of their small size or not collected because they were not questing during this season or were host-associated. Ticks of both species were significantly more abundant in grassland ($N=369$) than in forest (6) or manioc gardens (14) ($\chi^2=320.97$, df$=2$, $p<2\times10^{-16}$). Interestingly, for both species and across all habitat types, the morning collection period was more productive (244 ticks) compared with afternoon (137) ($\chi^2=14.9314$, df$=1$, $p=0.0001$).

Risks of tick-borne disease for villagers and animals are changing as a result of many of the responses outlined in Figure 5.2. Our finding of relatively few forest ticks is suggestive of the "kind forest" hypothesis (Figure 5.1b). When villagers hunt and livestock graze in forest (as they commonly do at night), they eliminate the buffer between them and sylvatic disease cycles, which has the potential to increase risk (Figure 5.1b). The manioc and peanut gardens serve as important sources of nutrition and are accessed by walking across the grasslands. In the gardens, nesting and molting sites for ticks are heavily disturbed. However, high tick-borne disease risk occurs in grassland, consistent with increased habitat quality for ticks and a shift to tick species more directly associated with animals or animal nests. It is not known if the low-intensity burns performed in DRC are capable of killing ticks directly. The altered microenvironment temperature and relative humidity probably did impact the ticks, as evidenced by the majority of ticks being collected in the morning. Moreover, extensive disruption of plant and animal communities as a result of fire management of savanna likely promotes increased tick-borne disease risk in this region. Clearly, more data to understand detailed mechanisms behind this intense grassland infestation are needed.

Tanzania

Tick-borne *Babesia bicornis*, other *Babesia, Theileria*, and *Ehrlichia* species, and direct tick-induced poor condition were observed, along with co-infecting viruses, among African buffalo (*Syncerus caffer*), wildebeest (*Connochaetes taurinus*), lions (*Panthera leo*), and black rhinoceros (*Diceros bicornis*) in the Ngorongoro Crater, Tanzania, between 2000 and 2001 (Fyumagwa et al. 2007). Investigators and agency authorities concluded that changes in local habitat conditions allowed for high tick densities. Removal of cattle grazing, together with 27 years of fire suppression, allowed tall grass species to predominate, resulting in improved tick habitat. Further, these tall grass species are preferred by buffalo that support large tick loads.

Controlled burns were approved and investigators were able to measure tick densities and vegetation characteristics before and for several years after the burns, which began in September 2001 and continued biennially. Prior to the burn, adult ticks were active from March–June in the wet season, while far more immature ticks were collected during dry conditions in September and October. If sites had short grass or were heavily grazed, they tended to have far fewer adult ticks and somewhat fewer immature ticks. On re-evaluation after one year, control areas had similar numbers of ticks as before the burn, while burned areas had virtually no adult ticks, an order of magnitude fewer immature ticks, and from one-fifth to one-third the grass height and fuel load of unburned sites. Despite intense grazing by a diversity of large herbivores, the authors noted that community composition may not have been optimal to control tall grass and that pastoralists had used fire as part of their management regime until approximately 30 years prior. The authors were not able to assess how the community would return to its pre-burn state because fire was planned to re-occur regularly and they did not report how the fires impacted risk of tick-borne disease specifically, but such a dramatic reduction in ticks would presumably have reduced disease burden on animals as well.

Brazilian Spotted Fever

The reservoir and vector of spotted fever rickettsiosis in South America (*Amblyomma cajennense sensu lato*) caused by *Rickettsia rickettsii* is favored by deforestation (Szabó et al. 2007). Transstadial (among developmental stages) and transovarial (from adult to offspring through eggs) transmission of *R. rickettsii* by *A. cajennense* maintains the pathogen, but this alone may not support the pathogen indefinitely (Soares et al. 2012). The proposed mechanism for the increase of *A. cajennense* outside of intact rain forest is that, while more specialized tick species will decrease in abundance with that of their preferred host, *A. cajennense* exploits a wide range of large mammals and thrives in a livestock-dominated and human-frequented landscape. Moreover, a primary host of *A. cajennense* is the capybara (*Hydrochoerus hydrochaeris*) (Perez et al. 2008), a species which responds positively to anthropogenic disturbance (de Barros Ferraz et al. 2007; Campos-Krauer and Wisely 2011). Thus, disturbance is likely to increase the risk of tick-borne disease following the "kind forest" hypothesis (Figure 5.1b).

Lyme Disease

In the eastern United States, Lyme disease, caused by *Borrelia burgorferi*, is vectored by the deer tick (*I. scapularis*). Larval *I. scapularis* do not acquire the pathogen from the prior generation and must obtain it from their larval or nymphal bloodmeal host before they can transmit to another host. In the mixed deciduous forests of the northeastern United States, immature *I. scapularis* exhibit a strong host preference for white-footed mice (*Peromyscus leucopus*). Adult ticks depend on white-tailed deer (*Odocoileus virginianus*) and other large mammals for blood meals.

The reforestation and subsequent fragmentation of the region created ideal conditions for white-footed mice, white-tailed deer, and the emergence of Lyme disease in the late twentieth century. As the center of American agriculture relocated to the plains, northeastern farmland was abandoned and returned to mixed hardwood forest. The deer population responded by growing exponentially (Côté et al. 2004). As browsers on low, woody vegetation, deer prefer forest edges that support such vegetation that is absent in dense forest. White-footed mice are densest in small forest patches, potentially owing to greater structural complexity provided by a higher edge-to-interior ratio (Nupp and Swihart 1996, 1998; Anderson et al. 2003). Their overlap with large tick populations supported by deer and high reservoir competence for *B. burgdorferi* likely led to the emergence of Lyme disease (Spielman 1994). This mechanism is in line with the "kind forest" hypothesis, where the undisturbed forest limits populations of important tick and pathogen hosts (Figure 5.1b). In addition to forest change, other anthropogenic factors such as climate change may influence the Lyme disease system (DellaSala and Hanson 2015).

Crimean-Congo Hemorrhagic Fever

A similar ecological cascade is thought to have driven the recent emergence of Crimean-Congo hemorrhagic fever (CCHF) in Turkey and Bulgaria. The incidence of this viral tick-borne disease increased 100-fold in Turkey from 2003 to 2008. A trend toward increased habitat fragmentation coinciding with foci of new cases suggests that anthropogenic disturbance contributed to the epidemic (Estrada-Peña et al. 2010). In neighboring Bulgaria, a higher incidence of CCHF occurs in highly fragmented habitats (Vescio et al. 2012). The mechanism behind these correlations has not yet been fully elucidated. The preferred habitats of the *Hyalomma* spp. ticks that vector CCHF are shrub- and grass-dominated systems, which may be more common in disturbed areas (Messina et al. 2015).

Tick-borne Encephalitis

Coarse-scale data on tick-borne encephalitis virus (TBEV) and forest management have been produced for municipalities in Latvia. Within rural parishes, the intensity of forest felling was positively associated with TBE incidence over a four-year period (Vanwambeke et al. 2010). The authors also quantified the edginess of forest fragments using a shape index: parishes with high incidence of TBE also had among the highest mean shape indices, identifying the amount of forest edge habitat as a risk factor (Vanwambeke et al. 2010). These associations may be related to the "isolated forest" hypothesis (Figure 5.1b), as both logging and edginess can bring humans into contact with forest-dwelling ticks.

Rizzoli and colleagues approached a similar question for TBE in northern Italy. TBE-positive provinces had more high-canopy forest compared to coppice (short woodlands frequently harvested for firewood) and higher roe deer (*Capreolus capreolus*) densities (Rizzoli et al. 2009). High forest is thought to support denser populations of yellow-necked mice (*Apodemus flavicollis*), a species characteristic of mature forest and the primary reservoir host of TBEV in Italy. As is the case with Lyme disease, high deer populations are thought to be a precursor for TBE emergence due to their capacity to host large numbers of adult ticks. The connection between yellow-necked mice and undisturbed forest supports the "cruel forest" hypothesis (Figure 5.1a).

The Missing Pieces

Our current understanding of anthropogenic influences on tick-borne disease derives from a variety of studies and observations, typically at very broad spatial scales, although occasionally at very fine scales. Many factors affecting tick-borne disease outcomes following disturbance have received sparse attention: across spatial scales, habitat types, disturbance intensity, disturbance "shape" (i.e. whether the disturbance occurred in one large area or occurred in smaller, possibly connected patches intermixed with non-disturbed areas), and for remote sites versus those adjacent to already-disturbed sites (Table 5.1). Most acutely, research is lacking on fine scales for timber harvesting, any scale for high-intensity wildfire, and detailed mechanisms that link disturbance through a causal pathway to tick and disease outcomes. Importantly, the risk of tick-borne disease can vary greatly in space with pronounced hot spots and cold spots (Dobson et al. 2011; Fedorova et al. 2014). Below, we consider other studies on ecosystem change and ticks that could address some mechanisms for how disturbance can directly and indirectly modify tick communities. These include changes to the plant community, host-community composition, host behavior, host distribution, and microclimate conditions. Putting these data together with the current literature can expand our understanding and help establish a baseline for future studies to improve our power to predict and mitigate outcomes.

Microclimate

Harvesting or fire-induced removal of canopy trees leads to higher temperature and lower relative humidity at the forest floor (Brooks and Kyker-Snowman 2008). Higher levels of sunlight allow for the recruitment of a variety of shade-intolerant shrub and tree species (Fredericksen et al. 1999). This shift in the plant community may change the type, quality, and quantity of resources available to both the tick and host communities. In addition, the collateral effects (e.g. skid trails, road building, lopping of slash) of timber harvesting can lead to soil compaction (Williamson and Neilsen 2000) and a decrease in the amount of woody debris (Tinker and Knight 2000). The movement of stems through the forest can disrupt litter and understory vegetation, modifying the habitat of ticks and their ground-dwelling hosts.

A decrease in tick population density is expected to follow any significant canopy-opening disturbance (Figure 5.2). A shift in community composition towards ticks with higher tolerance for desiccation is also possible. Lower humidity, higher temperatures, and disturbance of the litter should decrease tick survival. Movement of hosts away from disturbed areas will decrease the feeding success of the standing tick population. Different species of lizards, birds and mammals are expected to respond differently to the loss of trees. Each group has a range of preferences along a full spectrum of canopy openness. An expected decrease in tick density due to decreased tick survival may be counterbalanced by changes in host behavior (Figure 5.2). Thus, any prediction of a change in disease risk is nearly impossible to support without studies that specifically monitor ticks before and after disturbance.

Small Mammals

A large number of studies have quantified the response of mammal communities to logging operations and fire. A meta-analysis of small mammal response to timber harvests in North America showed that most species respond positively to harvest such that abundances in disturbed areas exceed pre-harvest and undisturbed sites (Zwolak 2009). An exception is the red-backed vole (*Myodes gapperi*) which becomes less abundant following logging. Other species, such as chipmunks (*Tamias* spp.) which tend to favor older forests with large amounts of downed woody debris, may decline in abundance following timber harvesting and fire. This can be significant because eastern chipmunks (*T. striatus*), redwood chipmunks (*T. ochrogenys*), and Allen's chipmunks (*T. senex*) are all capable of serving as reservoirs for tick-borne pathogens (Slajchert et al. 1997; Foley et al. 2008; Nieto and Foley 2009). Like timber harvesting, the impact of fire on small mammals shows a complete array of negative, neutral and positive responses (Fontaine and Kennedy 2012). Changes in small mammal densities after disturbance can lead to changes in predator behavior (Fuller and Harrison 2005). Mammal relative abundance may also be influenced by diet. Tree felling can add a large volume of seeds to the forest floor, potentially creating more favorable habitat for granivores (Gunther et al. 1983; Fisher and Wilkinson 2005). An overall higher density of small mammals should benefit immature, host-seeking ticks, increasing the chance of survival to the next stage (Figure 5.2). The outcome of tick-borne disease risk is ultimately dependent on how reservoir hosts respond, which vary in their sensitivity to disturbance among the many small mammals that maintain tick-borne pathogens.

Large Mammals

Many medically important tick species depend on ungulates, especially deer, as a source of adult blood meals. Both observational and manipulative studies confirm that tick populations are often limited by the relative abundance of deer. Thus, any change in deer abundance or behavior following harvest or fire should help to dictate the response of the tick community to disturbance (Figure 5.2). The removal of white-tailed deer from a site in Massachusetts gradually reduced tick populations as adults experienced reduced feeding and reproductive success (Wilson et al. 1988). A six-fold increase in lone star tick density after prescribed burning in Missouri was linked to an increase in local deer density (Allan 2009). Since post-logging plant recruitment is expected to be similar to that of a post-fire community, deer are likely to bring ticks into harvested areas as well and restore the tick community (Figure 5.2). The rise in prevalence of four lone star tick-vectored pathogens was thought to be associated with increased white-tailed deer abundance across the eastern half of North America (Paddock and Yabsley 2007). At high density, deer can limit natural forest regeneration through excessive herbivory of understory vegetation (Miller et al. 2009). Thus, deer may support high tick populations for prolonged periods after disturbance.

Plant Community

Vegetation changes following logging and fire may modify habitat and food sources for ticks and their hosts. Response of the tick community to disturbance is likely to be driven by the response of hosts, which in turn should be strongly influenced by the plant community (Figure 5.2). Logging and other disturbances can create opportunities for the invasion of nonnative plants. In a controlled experiment, invasion of Japanese stiltgrass (*Microstegium vimineum*) was shown to decrease the survival of both lone star and American dog ticks (*Dermacentor variabilis*). Reduced survival was attributed to hotter and drier conditions within the invaded plots compared with uninvaded controls (Civitello et al. 2008). Along the United States–Mexico border, invasive giant reed (*Arundo donax*) creates conditions that support the survival of cattle ticks (*Rhipicephalus* spp.), the vector of bovine babesiosis (Racelis et al. 2012). Eradication of an invasive honeysuckle (*Lonicera maackii*) in Missouri resulted in reduced risk of lone star tick bites on people (Allan et al. 2010). The purported mechanism is that dense honeysuckle stands attract white-tailed deer, which support large tick populations. Manual removal of the invasive shrub reduced deer and tick density. In northern Argentina, widespread deforestation has created new pasture land for cattle and facilitated the expansion of non-native grasses (Nava et al. 2013). There were no significant differences in cattle tick (*Rhipicephalus microplus*) survival or reproductive timing between nonnative grasslands and the natural forested areas. Microclimatic conditions were similar between the two habitats, indicating that a land use change away from forest does not necessarily improve or diminish tick habitat.

Studies that connect tree pathogens and tick-borne disease are of interest to our central question, i.e. the influence of human-caused disturbance on disease, because they examine what happens following the selective loss of canopy trees. Below-ground herbivory by phytopathogenic *Leptographium* spp. fungi initiates a cascade in Wisconsin red pine (*Pinus resinosa*) forests, leading to increased tick density (Coyle et al. 2013). The fungus enables infestation by bark beetles, leading to tree death and opening gaps in the canopy. The gaps initiate recruitment of shrubby vegetation favored by deer and rodents. The higher host density supports a denser population of ticks and may lead to heightened disease risk (Coyle et al. 2013). In California, oak (*Quercus* spp.) mortality from sudden oak death (*Phytophthora ramorum*) leads to a more-open canopy and better habitat for deer mice (*Peromyscus maniculatus*) and western fence lizards (*Sceloporus occidentalis*). Increased density of these animals increases the density of nymphal ticks, likely due to increased larval feeding success (Swei et al. 2011a, 2012).

The Human Component

Beyond changes in the tick community following disturbance, any change in disease risk will be determined at the tick–human interface. Forestry and fire bring loggers, managers, and scientists into the forest, increasing their risk of exposure to ticks and pathogens. Forestry workers around the world are exposed to and suffer from tick-borne diseases at an increased rate compared with the general population (Covert and Langley 2002).

Human risk of tick-borne disease may vary with the type of disturbance. Even-aged forest management (i.e., clear-cutting) typically employs more machinery, thus reducing human contact with ticks. Uneven-aged management schemes rely more on human labor, potentially increasing the risk of these workers (Wolfe et al. 2005). A selective approach to timber harvesting is increasingly preferred as a strategy to target high-value species, save costs, preserve wildlife habitat, and maintain ecosystem services (Wolfe et al. 2005). Prescribed burning involves significant human exposure before, during, and after fire. Firefighting, fire suppression (including prescribed burning), and post-fire management all bring workers into close contact with ticks and pathogens. This is in contrast to many high-intensity fires, where human exposure to tick-borne disease is low.

The post-disturbance landscape may include features that modify human risk of acquiring a tick-borne disease. Namely, the construction of logging and fire roads can make otherwise inaccessible forested areas available to people for recreation, hunting, and gathering (Figure 5.1b) (Wolfe et al. 2005). Further, roads can serve as a conduit, spreading hosts, ticks, and pathogens into new areas. The creation of forest fragments and increased edge by road construction can drive the emergence of some tick-borne diseases, as discussed above. Specific risks are also affected by how attractive the post-disturbance landscape is for recreation, resource harvesting, and other activities.

Conclusions

Our changing planet presents myriad challenges for society. Among them is the threat of emerging and re-emerging spillover pathogens that threaten humans and our domestic animals. In this chapter, we have outlined the case for two disturbances and their implications for tick-borne disease risk. Although a generalizable theory for the relationship between all parasites and all disturbance types would be convenient, we find in our synthesis that many realities will prevent this.

Exposure to pathogens is the ultimate outcome of interest for disease researchers. Tick-borne disease risk is most often quantified as the estimated density of infected ticks. This value is dependent on the complex interactions among climate, vegetation, vertebrate hosts, and ticks (Figure 5.2). For a given pathogen and ecosystem disturbance, any number of hypotheses can be generated to predict a change in disease risk (Figure 5.1). We hope this framework influences future work on the health impacts of ecosystem disturbances.

A complete understanding is hampered by a lack of before-after-control (BACI) studies. In examining both fire and timber harvesting, we find conflicting possibilities as to how these disturbances affect ticks. We also frequently encounter that antecendent conditions or a change in scale or intensity of the disturbance fundamentally alter the questions we can ask and answer (Table 5.1). However, by examining each step along a causal pathway from disturbance to outcome in the tick-borne disease system, we can begin to build a set of predictions for what the response would be to disturbance in microenvironments, plant communities, large and small vertebrate communities, and ticks. Further research should not only expand the scope of their pre- and post-disturbance observations on tick-borne disease but also explicitly incorporate metrics for intensity, homogeneity, shape, spatial and temporal scale, and attributes of nearby habitat that will influence post-disturbance succession (Table 5.1). Realistically, however, ecological contingency is critically important in vector-borne disease response to disturbance and accurate prediction will be elusive. Nevertheless, building a database of information such as described here will allow us to organize and synthesize our understanding of changes in tick-borne disease risk across both space and time in our increasingly disturbed landscapes.

Acknowledgments

We thank Sarah Woodworth for leading the work in DRC, the people of Kikongo and Kimbuma, DRC, for access to their land and collecting assistance, and Walter Boyce for input during project design. Timothy Chapman provided assistance in DRC as guide, translator, and research assistant. Work in the DRC was funded by the UC Davis School of Veterinary Medicine STAR program and the Center for Vector-Borne Disease.

References

Abatzoglou, J. T., and A. P. Williams. 2016. Impact of anthropogenic climate change on wildfire across western US forests. *Proceedings of the National Academy of Sciences* 113: 11770–11775.

Allan, B. F. 2009. Influence of prescribed burns on the abundance of *Amblyomma americanum* (Acari: Ixodidae) in the Missouri Ozarks. *Journal of Medical Entomology* 46: 1030–1036.

Allan, B. F., H. P. Dutra, L. S. Goessling, K. Barnett, J. M. Chase, R. J. Marquis, G. Pang, G. A. Storch, R. E. Thach, and J. L. Orrock. 2010. Invasive honey-suckle eradication reduces tick-borne disease risk by altering host dynam-ics. *Proceedings of the National Academy of Sciences* 107: 18523–18527.

Allan, B. F., F. Keesing, and R. S. Ostfeld. 2003. Effect of forest fragmentation on Lyme disease risk. *Conservation Biology* 17: 267–272.

Anderson, C. S., A. B. Cady, and D. B. Meikle. 2003. Effects of vegetation struc-ture and edge habitat on the density and distribution of white-footed mice (*Peromyscus leucopus*) in small and large forest patches. *Canadian Journal of Zoology* 81: 897–904.

Balch, J. K., B. A. Bradley, J. T. Abatzoglou, R. C. Nagy, E. J. Fusco, and S. Mahood. 2017. Human-started wildfires expand the fire niche across the United States. *Proceedings of the National Academy of Sciences* 114: 2946–2951.

Bertrand, M. R., and M. L. Wilson. 1996. Microclimate-dependent survival of unfed adult *Ixodes scapularis* (Acari: Ixodidae) in nature: life cycle and study design implications. *Journal of Medical Entomology* 33: 619–627.

Brooks, R. T., and T. D. Kyker-Snowman. 2008. Forest floor temperature and rela-tive humidity following timber harvesting in southern New England, USA. *Forest Ecology and Management* 254: 65–73.

Brown, R. N., R. Lane, and D. T. Dennis. 2005. Geographic distributions of tick-borne diseases and their vectors. In J. L. Goodman, D. T. Dennis, and D. E. Sonenshine, editors, *Tick-Borne Diseases of Humans*, 363–391. Washington, D.C.: ASM Press.

Burridge, M. J., L.-A. Simmons, and S. A. Allan. 2000. Introduction of potential heartwater vectors and other exotic ticks into Florida on imported reptiles. *Journal of Parasitology* 86: 700–704.

Burt, F., D. Spencer, P. Leman, B. Patterson, and R. Swanepoel. 1996. Investigation of tick-borne viruses as pathogens of humans in South Africa and evidence of Dugbe virus infection in a patient with prolonged thrombocytopenia. *Epidemiology and Infection* 116: 353–361.

Campos-Krauer, J. M., and S. M. Wisely. 2011. Deforestation and cattle ranch-ing drive rapid range expansion of capybara in the Gran Chaco ecosystem. *Global Change Biology* 17: 206–218.

Civitello, D. J., J. Cohen, H. Fatima, N. T. Halstead, J. Liriano, T. A. McMahon, C. N. Ortega, E. L. Sauer, T. Sehgal, and S. Young. 2015. Biodiversity inhibits parasites: broad evidence for the dilution effect. *Proceedings of the National Academy of Sciences* 112: 8667–8671.

Civitello, D. J., S. L. Flory, and K. Clay. 2008. Exotic grass invasion reduces survival of *Amblyomma americanum* and *Dermacentor variabilis* ticks (Acari: Ixodidae). *Journal of Medical Entomology* 45: 867–872.

Côté, S. D., T. P. Rooney, J.-P. Tremblay, C. Dussault, and D. M. Waller. 2004. Ecological impacts of deer overabundance. *Annual Review of Ecology, Evolution, and Systematics* 35: 113–147.

Covert, D. J., and R. L. Langley. 2002. Infectious disease occurrence in forestry workers: a systematic review. *Journal of Agromedicine* 8: 95–111.

Coyle, D. R., M. W. Murphy, S. M. Paskewitz, J. L. Orrock, X. Lee, R. J. Murphy, M. A. McGeehin, and K. F. Raffa. 2013. Belowground herbivory in red pine stands initiates a cascade that increases abundance of Lyme disease vectors. *Forest Ecology and Management* 302: 354–362.

Cully, J. F. 1999. Lone star tick abundance, fire, and bison grazing in tallgrass prairie. *Journal of Range Management* 52: 139–144.

Cumming, G. S. 2000. Using habitat models to map diversity: pan-African species richness of ticks (Acari: Ixodida). *Journal of Biogeography* 27: 425–440.

Davidson, W. R., D. A. Siefken, and L. H. Creekmore. 1994. Influence of annual and biennial prescribed burning during March on the abundance of *Amblyomma americanum* (Acari: Ixodidae) in central Georgia. *Journal of Medical Entomology* 31: 72–81.

de Barros Ferraz, K. M. P., S. F. de Barros Ferraz, J. R. Moreira, H. T. Z. Couto, and L. M. Verdade. 2007. Capybara (*Hydrochoerus hydrochaeris*) distribution in agroecosystems: a cross-scale habitat analysis. *Journal of Biogeography* 34: 223–230.

de Castro, J. J. 1997. Sustainable tick and tickborne disease control in livestock improvement in developing countries. *Veterinary Parasitology* 71: 77–97.

DellaSala, D. A., and C. T. Hanson. 2015. *The Ecological Importance of Mixed-severity Fires: Nature's Phoenix*. Elsevier, Amsterdam.

Dingler, R. J., S. A. Wright, A. Donahue, P. Madeco, and J. Foley. 2014. Surveillance for *Ixodes pacificus* and the tick-borne pathogens *Anaplasma phagocytophilum* and *Borrelia burgdorferi* in birds from California's Inner Coast Range. *Ticks and Tick-borne Diseases* 5: 436–445.

Dobson, A. D., J. L. Taylor, and S. E. Randolph. 2011. Tick (*Ixodes ricinus*) abundance and seasonality at recreational sites in the UK: hazards in relation to fine-scale habitat types revealed by complementary sampling methods. *Ticks and Tick-borne Diseases* 2: 67–74.

Drew, M. L., W. Samuel, G. Lukiwski, and J. Willman. 1985. An evaluation of burning for control of winter ticks, *Dermacentor albipictus*, in central Alberta. *Journal of Wildlife Diseases* 21: 313–315.

Estrada-Peña, A., Z. Vatansever, A. Gargili, and Ö. Ergönul. 2010. The trend towards habitat fragmentation is the key factor driving the spread of Crimean-Congo haemorrhagic fever. *Epidemiology and Infection* 138: 1194–1203.

FAO. 2015. *Global Forest Resources Assessment 2015*. Rome.

Fedorova, N., J. E. Kleinjan, D. James, L. T. Hui, H. Peeters, and R. S. Lane. 2014. Remarkable diversity of tick or mammalian-associated Borreliae in the metropolitan San Francisco Bay Area, California. *Ticks and Tick-borne Diseases* 5: 951–961.

Fisher, J. T., and L. Wilkinson. 2005. The response of mammals to forest fire and timber harvest in the North American boreal forest. *Mammal Review* 35: 51–81.

Foley, J., S. B. Clueit, and R. Brown. 2008. Differential exposure to *Anaplasma phagocytophilum* in rodent species in northern California. *Vector-Borne and Zoonotic Disease* 8: 49–55.

Fontaine, J. B., and P. L. Kennedy. 2012. Meta-analysis of avian and small-mammal response to fire severity and fire surrogate treatments in US fire-prone forests. *Ecological Applications* 22: 1547–1561.

Fredericksen, T. S., B. D. Ross, W. Hoffman, M. L. Morrison, J. Beyea, B. N. Johnson, M. B. Lester, and E. Ross. 1999. Short-term understory plant community responses to timber-harvesting intensity on non-industrial private forestlands in Pennsylvania. *Forest Ecology and Management* 116: 129–139.

Fuller, A. K., and D. J. Harrison. 2005. Influence of partial timber harvesting on American martens in north-central Maine. *Journal of Wildlife Management* 69: 710–722.

Fyumagwa, R. D., V. Runyoro, I. G. Horak, and R. Hoare. 2007. Ecology and control of ticks as disease vectors in wildlife of the Ngorongoro Crater, Tanzania. *South African Journal of Wildlife Research* 37: 79–90.

Gottdenker, N. L., D. G. Streicker, C. L. Faust, and C. Carroll. 2014. Anthropogenic land use change and infectious diseases: a review of the evidence. *EcoHealth* 11: 619–632.

Gunther, P. M., B. S. Horn, and G. Babb. 1983. Small mammal populations and food selection in relation to timber harvest practices in the western Cascade Mountains. *Northwest Science* 57: 32–44.

Harrod, R., and S. Reichard. 2000. Fire and invasive species within the temperate and boreal coniferous forests of western North America. *Proceedings of the Invasive Species Workshop*: 95–101.

Hoch, A., P. J. Semtner, R. W. Barker, and J. A. Hair. 1972. Preliminary observations on controlled burning for lone star tick (Acarina: Ixodidae) control in woodlots. *Journal of Medical Entomology* 9: 446–451.

Hoogstraal, H. 1971. Identity, hosts, and distribution of *Haemaphysalis* (*Rhipistoma*) *canestrinii* (Supino)(resurrected), the postulated Asian progenitor of the African *leachi* complex (Ixodoidea: Ixodidae). *Journal of Parasitology* 57: 161–172.

Hulbert, L. C. 1986. Fire effects on tallgrass prairie. In G. K. Clambey and R. H. Pemble, editors, *Proceedings of the Ninth North American Prairie Conference*, 138–l142. Fargo, North Dakota: Tri-College University Center for Environmental Studie.

Iloweka, E. M. 2004. The deforestation of rural areas in the Lower Congo province. *Environmental Monitoring and Assessment* 99: 245–250.

Jefferies, R., U. M. Ryan, C. J. Muhlnickel, and P. J. Irwin. 2009. Two species of canine *Babesia* in Australia: detection and characterization by PCR. *Journal of Parasitology* 89: 409–412.

Lane, R. S., and G. Quistad. 1998. Borreliacidal factor in the blood of the western fence lizard (*Sceloporus occidentalis*). *Journal of Parasitology* 84: 29–34.

Lorimer, C. G., D. J. Porter, M. A. Madej, J. D. Stuart, S. D. Veirs, S. P. Norman, K. L. O'Hara, and W. J. Libby. 2009. Presettlement and modern disturbance regimes in coast redwood forests: implications for the conservation of old-growth stands. *Forest Ecology and Management* 258: 1038–1054.

Malcolm, J. R., and J. C. Ray. 2000. Influence of timber extraction routes on Central African small-mammal communities, forest structure, and tree diversity. *Conservation Biology* 14: 1623–1638.

Mather, T. N., D. C. Duffy, and S. R. Campbell. 1993. An unexpected result from burning vegetation to reduce Lyme disease transmission risks. *Journal of Medical Entomology* 30: 642–645.

Matthysse, J. G., and M. H. Colbo. 1987. The ixodid ticks of Uganda, together with species pertinent to Uganda because of their present known distribution. Entomological Society of America.

McCoy, K. D., E. Léger, and M. Dietrich. 2013. Host specialization in ticks and transmission of tick-borne diseases: a review. *Frontiers in Cellular and Infection Microbiology* 3: 57.

Messina, J. P., D. M. Pigott, N. Golding, K. A. Duda, J. S. Brownstein, D. J. Weiss, H. Gibson, T. P. Robinson, M. Gilbert, and G. W. Wint. 2015. The global distribution of Crimean-Congo hemorrhagic fever. *Transactions of The Royal Society of Tropical Medicine and Hygiene*: trv050.

Miller, B. F., T. A. Campbell, B. R. Laseter, W. M. Ford, and K. V. Miller. 2009. White-tailed deer herbivory and timber harvesting rates: implications for regeneration success. *Forest Ecology and Management* 258: 1067–1072.

MODIS Rapid Response Team. 2010. Fires in Democratic Republic of Congo, http://earthobservatory.nasa.gov/naturalhazards/view.php?id=43950.

Molyneux, D. H. 2003. Common themes in changing vector-borne disease scenarios. *Transactions of The Royal Society of Tropical Medicine and Hygiene* 97: 129–132.

Nava, S., M. Mastropaolo, A. A. Guglielmone, and A. J. Mangold. 2013. Effect of deforestation and introduction of exotic grasses as livestock forage on the population dynamics of the cattle tick *Rhipicephalus (Boophilus) microplus* (Acari: Ixodidae) in Northern Argentina. *Research in Veterinary Science* 95: 1046–1054.

Nieto, N. C., and J. E. Foley. 2009. Reservoir competence of the redwood chipmunk (*Tamias ochrogenys*) for *Anaplasma phagocytophilum*. *Vector-Borne and Zoonotic Diseases* 9: 573–577.

Nupp, T. E., and R. K. Swihart. 1996. Effect of forest patch area on population attributes of white-footed mice (*Peromyscus leucopus*) in fragmented landscapes. *Canadian Journal of Zoology* 74: 467–472.

Nupp, T. E., and R. K. Swihart. 1998. Effects of forest fragmentation on population attributes of white-footed mice and eastern chipmunks. *Journal of Mammalogy* 79: 1234–1243.

Ogden, N., A. Maarouf, I. Barker, M. Bigras-Poulin, L. Lindsay, M. Morshed, C. O'Callaghan, F. Ramay, D. Waltner-Toews, and D. Charron. 2006. Climate change and the potential for range expansion of the Lyme disease vector *Ixodes scapularis* in Canada. *International Journal for Parasitology* 36: 63–70.

Olsen, B., D. C. Duffy, T. G. Jaenson, A. Gylfe, J. Bonnedahl, and S. Bergstrom. 1995. Transhemispheric exchange of Lyme disease spirochetes by seabirds. *Journal of Clinical Microbiology* 33: 3270–3274.

Owen, C. E., S. Bahrami, J. C. Malone, J. P. Callen, and C. L. Kulp-Shorten. 2006. African tick bite fever: a not-so-uncommon illness in international travelers. *Archives of Dermatology* 142: 1312–1314.

Paddock, C. D., and M. J. Yabsley. 2007. Ecological havoc, the rise of white-tailed deer, and the emergence of *Amblyomma americanum*-associated zoonoses in the United States. In *Wildlife and Emerging Zoonotic Diseases: The Biology, Circumstances and Consequences of Cross-species Transmission*, 289–324. Berlin: Springer.

Padgett, K., L. Casher, S. Stephens, and R. Lane. 2009. Effect of prescribed fire for tick control in California chaparral. *Journal of Medical Entomology* 46: 1138–1145.

Parola, P., C. D. Paddock, and D. Raoult. 2005. Tick-borne rickettsioses around the world: emerging diseases challenging old concepts. *Clinical Microbiology Reviews* 18: 719–756.

Perez, C. A., Á. F. d. Almeida, A. Almeida, V. H. B. d. Carvalho, D. d. C. Balestrin, M. S. Guimarães, J. C. Costa, L. A. Ramos, A. D. Arruda-Santos, and C. P.

Máximo-Espíndola. 2008. Ticks of genus *Amblyomma* (Acari: Ixodidae) and their relationship with hosts in endemic area for spotted fever in the state of São Paulo. *Revista Brasileira de Parasitologia Veterinária* 17: 210–217.

Racelis, A., R. Davey, J. Goolsby, A. P. De León, K. Varner, and R. Duhaime. 2012. Facilitative ecological interactions between invasive species: *Arundo donax* stands as favorable habitat for cattle ticks (Acari: Ixodidae) along the US–Mexico border. *Journal of Medical Entomology* 49: 410–417.

Randolph, S. E., and D. J. Rogers. 2000. Fragile transmission cycles of tick-borne encephalitis virus may be disrupted by predicted climate change. *Proceedings of the Royal Society of London B: Biological Sciences* 267: 1741–1744.

Rimoin, A. W., and P. J. Hotez. 2013. NTDs in the heart of darkness: the Democratic Republic of Congo's unknown burden of neglected tropical diseases. *PLoS Neglected Tropical Diseases* 7: e2118.

Rizzoli, A., H. C. Hauffe, V. Tagliapietra, M. Neteler, and R. Rosà. 2009. Forest structure and roe deer abundance predict tick-borne encephalitis risk in Italy. *PLoS One* 4: e4336.

Roberts, S. L., D. A. Kelt, J. W. van Wagtendonk, A. K. Miles, and M. D. Meyer. 2015. Effects of fire on small mammal communities in frequent-fire forests in California. *Journal of Mammalogy* 96: 107–119.

Rogers, A. J. 1955. The abundance of *Ixodes scapularis* Say as affected by burning. *Florida Entomologist* 38: 17–20.

Salkeld, D. J., and R. S. Lane. 2010. Community ecology and disease risk: lizards, squirrels, and the Lyme disease spirochete in California, USA. *Ecology* 91: 293–298.

Scasta, J. D. 2015. Fire and parasites: an under-recognized form of anthropogenic land use change and mechanism of disease exposure. *EcoHealth*: 1–6.

Schoennagel, T., J. K. Balch, H. Brenkert-Smith, P. E. Dennison, B. J. Harvey, M. A. Krawchuk, N. Mietkiewicz, P. Morgan, M. A. Moritz, and S. Rasker. 2017. Adapt to more wildfire in western North American forests as climate changes. *Proceedings of the National Academy of Sciences* 114: 4582–4590.

Slajchert, T., U. D. Kitron, C. J. Jones, and A. Mannelli. 1997. Role of the eastern chipmunk (*Tamias striatus*) in the epizootiology of Lyme borreliosis in north-western Illinois, USA. *Journal of Wildlife Diseases* 33: 40–46.

Soares, J., H. Soares, A. Barbieri, and M. Labruna. 2012. Experimental infection of the tick *Amblyomma cajennense*, Cayenne tick, with *Rickettsia rickettsii*, the agent of Rocky Mountain spotted fever. *Medical and Veterinary Entomology* 26: 139–151.

Spielman, A. 1994. The Emergence of Lyme disease and human babesiosis in a changing environment. *Annals of the New York Academy of Science* 740: 146–156.

Stafford III, K. C., J. S. Ward, and L. A. Magnarelli. 1998. Impact of controlled burns on the abundance of *Ixodes scapularis* (Acari: Ixodidae). *Journal of Medical Entomology* 35: 510–513.

Swei, A., C. J. Briggs, R. S. Lane, and R. S. Ostfeld. 2012. Impacts of an introduced forest pathogen on the risk of Lyme disease in California. *Vector-Borne and Zoonotic Diseases* 12: 623–632.

Swei, A., R. S. Ostfeld, R. S. Lane, and C. J. Briggs. 2011a. Effects of an invasive forest pathogen on abundance of ticks and their vertebrate hosts in a California Lyme disease focus. *Oecologia* 166: 91–100.

Swei, A., R. S. Ostfeld, R. S. Lane, and C. J. Briggs. 2011b. Impact of the experimental removal of lizards on Lyme disease risk. *Proceedings of the Royal Society of London B: Biological Sciences* 278: 2970–2978.

Szabó, M. P. J., M. M. M. Olegário, and A. L. Q. Santos. 2007. Tick fauna from two locations in the Brazilian savannah. *Experimental and Applied Acarology* 43: 73–84.

Tinker, D. B., and D. H. Knight. 2000. Coarse woody debris following fire and logging in Wyoming lodgepole pine forests. *Ecosystems* 3: 472–483.

Vanwambeke, S. O., D. Šumilo, A. Bormane, E. F. Lambin, and S. E. Randolph. 2010. Landscape predictors of tick-borne encephalitis in Latvia: land cover, land use, and land ownership. *Vector-Borne and Zoonotic Diseases* 10: 497–506.

Vescio, F. M., L. Busani, L. Mughini-Gras, C. Khoury, L. Avellis, E. Taseva, G. Rezza, and I. Christova. 2012. Environmental correlates of Crimean-Congo haemorrhagic fever incidence in Bulgaria. *BMC Public Health* 12: 1116.

Williams, G. W. 2003. References on the American Indian use of fire in ecosystems. US Forest Service.

Williamson, J., and W. Neilsen. 2000. The influence of forest site on rate and extent of soil compaction and profile disturbance of skid trails during ground-based harvesting. *Canadian Journal of Forest Research* 30: 1196–1205.

Wilson, M. L., S. R. d. Telford, J. Piesman, and A. Spielman. 1988. Reduced abundance of immature *Ixodes dammini* (Acari: Ixodidae) following elimination of deer. *Journal of Medical Entomology* 25: 224–228.

Wolfe, N. D., P. Daszak, A. M. Kilpatrick, and D. S. Burke. 2005. Bushmeat hunting, deforestation, and prediction of zoonotic disease. *Emerging Infectious Diseases* 11: 1822–1827.

Zwolak, R. 2009. A meta-analysis of the effects of wildfire, clearcutting, and partial harvest on the abundance of North American small mammals. *Forest Ecology and Management* 258: 539–545.

Land-use Disturbance

Landscape Pattern, Resilience, and Recovery

chapter six

Microbes to Bobcats
Biological Refugia of Appalachian Reclaimed Coal Mines

Nicole Cavender and Suzanne Prange

Mining and Reclamation in Appalachia

The Appalachian region is one of the most biologically diverse areas in the temperate world (Ricketts et al. 1999), and it is considered an ecological "hotspot" (Stein et al., 2000.) This ecoregion provides important ecosystem services such as storing carbon and maintaining watershed and water quality (Zipper et al. 2011). Despite its ecological value, Appalachia is greatly impacted by anthropogenic disturbance, ranging from agriculture to timber harvest to surface mining (Gragson and Bolstad 2006).

Since the early part of the twentieth century, surface mining has been changing the Appalachian landscape (Yarnell 1998). Although over 600,000 hectares of Appalachian forests were mined between 1978 and 2009 (USOSM 2010), it is estimated that an additional 10,000 hectares are mined annually (Zipper et al. 2011). The process of surface mining alters the physical, chemical, and biological characteristics of the ecosystem (Jacobs 2005; Palmer et al. 2010). Surface mining eliminates existing vegetation, dramatically alters the soil profile and chemistry, and displaces or kills wildlife, among a myriad of other effects for both terrestrial and aquatic ecosystems (USDI 1979; Palmer et al. 2010; Lindberg et al. 2011).

Prior to the 1970s, mine reclamation was largely unregulated across the United States. This left many abandoned minelands throughout the landscape. Although evidence suggests that biodiversity will return to these abandoned minesites over time through succession (Skousen et al. 1994, 2006; Gorman et al. 2001), they are typically dominated by early-successional species or grasses (Skousen et al. 1994, 2006). Given that these young forests differ from undisturbed sites, and they typically do not receive active management, reintroducing rare species to increase diversity may not often be successful (Turner 2015).

To address the growing number of abandoned minesites and the increasing environmental degradation, the federal government enacted the Surface Mining Control and Reclamation Act (SMCRA) in 1977 (SMCRA 2006). These laws improved environmental quality, including issues related to erosion control, water quality, extreme pH, and land stability (Casselman et al. 2006). More-recent reclamation practices include The Forestry Reclamation Approach (FRA), which was developed as a set of reclamation practices for mineland to support timber production (Burger et al. 2005).

Despite the federally-mandated improvements from SMCRA, the historic process of reclaiming mine sites created and led to additional ecological challenges. Grading equipment caused significant compaction of the soil, which resulted in low soil porosity, permeability, and moisture-holding capacity (Torbert and Burger 2000). Furthermore, although SMCRA required the site be restored, the site did not need to be restored to its original ecosystem. In most cases, the post-mining plan specified forage grasses and forbs for agricultural use rather than native forest (Holl 2002). Consequently, most surface-mined land in the Appalachian ecoregion was not reclaimed to forest (US GAO 2009), and native forests were replaced with exotic grasslands.

In North America, as well as globally, grasslands are one of the most threatened habitat types (Noss et al. 1995; Williams and Diebel 1996). The great majority of North America's grasslands have been converted to agricultural lands for crop and livestock production. Thus, although prairies are not the natural ecosystem of the Appalachian ecoregion, further restoration of SMCRS-reclaimed mineland to prairie could represent refugia for many species declining because of grassland habitat loss, while potentially allowing the land to recover, albeit slowly, to its native forest character. Given that prairie is now a critically endangered habitat, measures to utilize reclaimed minelands as refugia may help to offset the loss of species that depend on grassland habitat (Noss et. al 1995). Reclamation, when done properly and coupled with active management, can bring together anthropogenic and natural forces and be a valuable tool to mediate the large-scale past human disturbance of mining (Swab et al. 2017) and preserve the countless ecosystem services of the region (Zipper et al. 2011).

The Wilds

The Wilds is a large non-profit international wildlife ecotourism and conservation center located in the Appalachian ecoregion in southeastern Ohio (39°49′48″ N, 81°43′53″ W; Figure 6.1). An important part of its formation in 1984 was a gift of 3705 ha from the Central Ohio Coal Company, a subsidiary of the American Electric Power Company, one of the largest mining operations in the region. The land gifted to The Wilds has therefore experienced substantial disturbance from surface mining for coal,

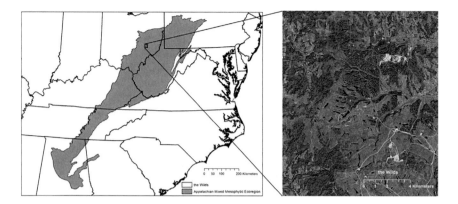

Figure 6.1 **(See color insert.)** Location of The Wilds within the Appalachian mixed mesophytic forest ecoregion in Ohio. Appalachian mixed mesophytic forest ecoregion map adapted from Olson et al. (2001).

although some remnant forest patches still remain. Over 90 percent of the landscape managed by The Wilds is in various stages of post-reclamation or restoration and represents a large, contiguous area of recovering reclaimed surface mineland. The vast land at The Wilds, the past conditions, and the dedication to conservation lends itself well to scientific study of recovering ecosystems and wildlife and the alternate pathways of recovery that can result after human disturbance.

Forested areas include 1552 ha reclaimed between the 1940s and 1970s. More recently, reclaimed areas consist of approximately 1885 ha of grasslands (Figure 6.2). In addition, there are more than 125 lakes and ponds (201 ha) and wetlands (55 ha) that include beaver ponds, marshes, vernal pools, and areas without open water. An important aspect to consider is that much of the landmass of The Wilds is characterized by the presence of limestone in the subsoil, which has a buffering effect to both soil pH and water even after mining and reclamation activities. The land has undergone drastic changes in ecosystem composition and functioning.

With the objective of ensuring healthy wildlife populations for global recovery initiatives and for public education, part of the acreage at The Wilds is used to manage and breed some of the most endangered animals on the planet such as white rhinos (*Ceratotherium simum*), scimitar horned oryx (*Oryx dammah*), and Sichuan takin (*Budorcas taxicolor tibetana*). Trained staff are dedicated to animal breeding and care and education tours are offered daily to the public.

Because of its history, size and dedication to conservation and education, The Wilds offers unique opportunities to explore central principles in ecology as they relate to community assembly, species interactions, biodiversity, succession, and response to ecological restoration. In 2002,

(a)

(b)

Figure 6.2 (a) and (b) Reclaimed exotic grasslands at The Wilds and forest patches.

The Wilds' staff initiated a formal Restoration Ecology program, and staff and other biologists began in earnest to analyze previous wildlife observations, conduct biological inventories, implement restoration management, and initiate ecologically-based field experiments. The Wilds' work to restore native habitats is the foundation of their conservation philosophy and practice, which is deeply rooted in a strong scientific approach (Bauman and Cavender 2011).

Information observed over decades from biological communities, including microbes (bacteria), cellular slime molds (dictyostelids), fungi, gastropods, vegetation, birds, pollinators, small mammals, and predators, have provided insight into biodiversity and its relation to multiple disturbances, habitat, recovery, succession, spatial distribution of habitats, management regimes, presence of refugia, and overall ecosystem functioning. Herein, we explore these datasets and insights to assess the ecological basis of changes in biological community composition of these taxa based on their distribution, richness, and abundance. Of particular interest is ecosystem functioning and wildlife recovery following mining and subsequent reclamation, and management practices that can promote stable, functioning ecosystems in these areas. What are these relationships between natural and anthropogenic influences, and how might we best intervene to promote the support of high biodiversity?

Flora and Fauna Post Reclamation

Vegetation

To examine the vegetation community in areas subjected to SMCRA-mandated reclamation followed by passive management, research staff at The Wilds evaluated vegetation species richness and community composition of reclaimed grassland areas that had been recovering from surface mining for >30 years. The results showed that pioneer seedlings were virtually absent and that little native recruitment had taken place (Cavender et al. 2014). Native plant species represented less than 2 percent of the area, whereas 98 percent consisted of introduced and naturalized plant species. The most abundant plants were those used in the original reclamation seed mix (Kentucky blue grass [*Poa pratensis*], tall fescue [*Festuca arundinacea*], Chinese lespedeza [*Lespedeza cuneate*], autumn olive [*Elaeagunus umbellata*], yellow sweet-clover [*Melilotus officinalis*], and bird's foot trefoil [*Lotus corniculatus*]), species selected for their ability to establish on nutrient-deficient and compacted soils (Torbert and Burger 2000). This dominant herbaceous cover likely inhibited the recruitment of native species, resulting in non-native grasslands in a state of arrested succession (Vaness and Wilson 2007). However, the cover allowed proliferation of the invasive shrub, autumn olive, which is successfully colonizing large swaths of these recovering mineland sites at The Wilds.

Invasive and naturalized species create barriers for native recruitment by altering nutrient cycles, water tables, and soil microbial communities from their natural condition (Vitousek 1990). Although previous studies have reported arrested succession in reclaimed mine areas (Holl 2002), it is noteworthy how little the plant community composition changed since the introduction of the exotic species, given the decades of time that had passed since the reclamation process ceased. It appears as though an ecological threshold was reached. Reasons for lack of succession to the native mixed-mesophytic forest community included: (1) lack of a native seed bank; (2) limited tree recruitment because of the distance to remnant forest edge; (3) soil compaction; and (4) dense exotic herbaceous growth. Others have also reported a lack of re-establishment by native tree species at similar SMCRA sites characterized by soil compaction and invasive ground cover (reviewed in Burger 2011). Even when sites were purposefully replanted with native trees, establishment and recolonization by native species often remain low (Simmons et al. 2008; Zipper et al. 2011). Thus, replanting based on reclamation regulations, followed by passive management, will not likely result in the re-establishment of the native plant community.

Without active management post-reclamation, nearly 80,000 hectares of land in Ohio will remain exotic grasslands in arrested succession (Kaster and Vimmerstedt 1996) or dense patches of invasive shrubland. One of the most invasive shrubs is autumn olive. This species was planted during reclamation; however, autumn olive has spread prolifically and control measures are difficult because of abundant fruit production and aggressive re-sprouting. The common management practices of prescribed burns and pruning are ineffective in reducing the spread of autumn olive. Prescribed burning does not limit the species' establishment. Pruning to the ground results in vigorous re-sprouting, a thicker multi-stemmed base, and, often, denser branches. To address this management issue, The Wilds research staff evaluated various control methods, including mechanical removal, foliar herbicide, and dormant-stem herbicide, for autumn olive. In 2007–2008, nine 200-m^2 study plots with three replications each were established. The foliar herbicide resulted in near-complete (98 percent) mortality of autumn olive. Dormant-stem herbicide caused 71 percent mortality, whereas mechanical control alone was not effective, achieving at only 15 percent mortality (Byrd and Cavender 2010). Although grasslands on reclaimed mineland are composed largely of exotic species, they can function as habitat for a diversity of native species as well, and the encroachment of invasive shrubs, such as autumn olive, constitutes a major threat (see Grassland birds below). As such, early control of invasive woody vegetation is imperative. The resulting landscape also suggests that avoiding the practice of utilizing non-native and invasive species in reclamation is a prudent measure. Substitution of native seed mixes into reclamation practices may provide additional benefits.

As an alternative to exotic cool-season grasslands, a diverse mix of prairie species may be useful in restoring greater ecological functioning to reclaimed minelands (Swab et al. 2017). The Wilds has converted >140 ha of reclaimed mineland to native prairie. Diverse prairie communities are stable and are associated with good soil development (Tilman and Downing 1994). As soil productivity, fertility, and organic matter increase (Tilman et al. 2006), diversity at many trophic levels may also increase, including the addition of pollinators and seed dispersers that may eventually lead to the succession to native forest (Cavender-Bares and Cavender 2011; Cusser and Goodell 2014).

Thorne and Cardina (2011) found that successful native prairie species on reclaimed calcareous mine soil at The Wilds were those that tolerated compacted soil and periodic anaerobic conditions (i.e., that occurred following heavy rainfall). Species that successfully established either rooted through compacted layers or had shallow roots and tolerated water stress. Eastern gamagrass (*Tripsacum dactyloides*), tall dropseed (*Sporobolus compositus*), buffalo grass (*Bouteloua dactyloides*), and western wheatgrass (*Pascopyrum smithii*) were the most successful grasses. Late-successional grasses, such as big bluestem (*Andropogon gerardii*), Indian grass (*Sorghastrum nutans*), and little bluestem (*Schizachyrium scoparium*) initially exhibited low establishment, but over time they were successful and persisted throughout the 2-year study period. Conversely, blue grama (*Bouteloua gracilis*) and sideoats grama (*Bouteloua curtipendula*) emerged the first year, but did not return in the second. The authors concluded this was caused by intolerance to anaerobic soil. Most areas required multiple herbicide treatments, and some areas required intensive invasive species control before seeding to achieve successful establishment (Cavender and Byrd 2007).

Methods of prairie species establishment have since been applied experimentally to reclamation of abandoned mined sites in Ohio in cooperation with the Ohio Department of Natural Resources, Division of Wildlife and The Wilds. Results from the study showed that utilizing prairie species seed mixes not only met reclamation standards but increased plant species richness, and indicated the likelihood of additional ecosystem benefits of greater soil microbial biomass, better pollinator resources, and greater potential value to wildlife (Swab et al. 2017).

Microbial Communities

Strip mining not only removes existing vegetation, it also alters the soil profile and reduces soil microbes, leaving organic carbon-poor soils (Mummey et al. 2002). Microorganisms in the soil are critical for decomposing organic residues, recycling soil nutrients, and enhancing the bioavailability of soil nutrients. They play a vital role in the redevelopment of vegetation on reclaimed mineland soils (Poncelet et al. 2014). Furthermore, microbial activities are fundamental in minelands to converting rock to soil (Chorover et al. 2007; Poncelet et al. 2014). However, geochemical change and soil re-establishment are slow processes, especially in passively managed reclaimed mineland.

The sites with overburden, the material above the coal seam, also referred to as spoil or waste, were compared to sites with undisturbed soil at The Wilds (Poncelet et al. 2014). Chemical and mineralogical characteristics of overburden samples were determined, as were microbial respiration rates. Results indicated that biogeochemical weathering of the overburden led to geochemical conditions and microbial communities that were similar to those of undisturbed soils. However, the recovery of the microbial community was still incomplete after 37 years.

Cellular Slime Molds

Despite a long history of heavy disturbance, The Wilds is home to a diversity of species across a wide range of taxonomic groups. One of the most diverse groups is that of cellular slime molds (CSM), the dictyostelids in the Kingdom Protista. Cellular slime molds are bacterivores that live in the surface layer of the soil, particularly forest soil where they have shelter and easy access to food. They have been studied in forest soils of Ohio since 1970, and various ecological (Cavender and Hopka 1986) and taxonomic and ecological (Cavender and Vadell 2006) studies have been published. An unglaciated, mixed mesophytic remnant forest patch, Butterfly Woods (BW), at The Wilds is a forest that is now surrounded by reclaimed mineland. Out of 119 sites sampled in Ohio, BW had the highest CSM diversity (Cavender and Cavender 2013). Out of the 25 species of slime molds that are found in Ohio, nineteen species were found in BW over a seven-year period. BW is the most-diverse CSM site found to date in any single temperate deciduous forest (<40 ha), worldwide. BW is believed to be optimal for CSM diversity for a number of reasons, the most surprising of which is its approximation to disturbance. Although, in general, disturbance can often have negative effects, some disturbance may actually increase CSM diversity. At least three species encountered, *Dictyostelium sphaero-cephalum*, *D. mucoroides*, and *D.giganteum*, prefer disturbed soils (Kauffman 1986; Hammer 1984). High species richness of CSM also contributed to herbaceous understory richness of BW as well as the fact that this region had not been glaciated. These results, along with other richness data, suggest that forest remnants are likely serving as refugia that enhance re-colonization of surrounding disturbed areas, as well as improving overall landscape connectivity.

Fungi

Like soil microbes, arbuscular mycorrhizal fungi (AMF) are essential to the recovery of reclaimed minelands (Levy and Cumming 2014). The symbiotic function of AMF is to supply plants with minerals, primarily phosphorus, in exchange for organic compounds needed by the fungi (Cavagnaro et al. 2005; Thorne et al. 2013). AMF may also affect plant community composition by forming beneficial associations with some species more readily than others (Gillespie and Allen 2006).

Thorne et al. (2013) examined AMF communities living symbiotically with cool-season, non-native grasses on reclaimed mineland at The Wilds. They determined the suitability of the AMF present for the establishment of native, warm-season grasses. Their results, based on root colonization assessment, indicated that AMF associated with reclaimed mineland soil would not likely limit the establishment of native, tallgrass prairie. Although the AMF in reclaimed mineland soil had been associated with exotic grasses for 30 years, apparently no host-specific relationship was established that would be an obstacle to native grassland restoration in this area. Poor soil condition in the reclaimed mineland might have actually promoted the growth of an AMF community that would benefit, rather than impede, native grassland species.

6.3.5 Gastropods

Terrestrial invertebrates are another taxonomic group that is affected by surface mine reclamation. Terrestrial snails are common, yet poorly studied, members of the Appalachian fauna. Watters et al. (2005) compared the terrestrial snail communities in a reclaimed area at The Wilds with that of a non-mined area in the Shawnee State Forest, also located in the Appalachian mixed mesophytic forest ecoregion in Ohio. The lowland woods of Shawnee shared many of the same species found in the wooded and marginal areas at The Wilds. The authors concluded that these areas at The Wilds were more humid, accumulated more leaf litter, and were thermally buffered. Upland areas at Shawnee State Forest were drier, less vegetated, and subject to thermal extremes. Species living in sparse fields and wood margins at Shawnee also inhabited exotic grasslands at The Wilds. Some of these species existed in the wooded uplands of Shawnee as well, but apparently required microhabitats found in dry uplands, wood margins, or grasslands to flourish. The authors speculated that the grassland species exist in the proper microhabitat in wooded areas and facultatively colonize grasslands and other open spaces as they become available. As observed in the CSM, remnant forests at The Wilds served as refugia for woodland snails from which surrounding areas could be repopulated. Furthermore, the conversion of woodland into grassland allowed some snails, once isolated in rare woodlands, to colonize these areas and become abundant.

The original snail assemblage at The Wilds has thus become two communities, a grassland and woodland fauna, representing an overall increase in diversity. Fifty species were recorded from Shawnee, whereas 60 were found at The Wilds. Twelve woodland species were found at both sites. In total, 32 woodland species were recorded at Shawnee and 24 at The Wilds, four grassland species occurred at Shawnee and 18 at The Wilds, and eight species occurred in both habitats at Shawnee, whereas

16 did so at The Wilds (Watters et al. 2005). The Chao 1 estimator, which predicts species richness from a sample, predicted 52 species for Shawnee State Forest and 78 for The Wilds, out of a total of 80 species known to occur in all of Ohio.

Lepidopterans and Other Pollinators

The Wilds harbors rich pollinator communities; it contains at least 94 bee species and 40 butterfly species. Planted native-prairie and meadow communities attract increasingly more butterflies, but proximity to remnant hardwood forest seems to be one of the most important factors for bee and butterfly richness. The forest likely provides nesting sites for some bee species, early-spring energy sources and host plants for Lepidoptera, and refugia from predators. Pollination is a functional relationship required for successful ecological restoration (Dixon 2009). Whereas soil microbes and AMF prepare the soil and contribute to plant growth, pollinators play vital roles in the maintenance and stability of plant communities (Kearns et al. 1998). Pollination mutualisms cannot be reestablished until all requirements of pollinators (e.g., feeding, nesting, and over-wintering areas) are met (Roulston and Goodell 2011).

In 2003, large-scale restoration efforts began at The Wilds to convert exotic, cool-season grasses to native, warm-season, nectar-producing grassland species. The two major goals of the restoration effort were to improve plant diversity and increase butterfly populations and diversity. Four years following restoration, perennial plants increased from 11 to 93 species representing 11 and 30 families, respectively. Furthermore, butterflies responded quickly and increased from an average of 653 butterflies recorded over a 23-week period to an average of 2138 butterflies recorded, a 227 percent increase. Species richness for butterflies likewise increased by 42 percent (Cavender-Bares and Cavender 2011).

In a related effort, Goodell et al. (2010) reported that the richness of bees (at the genus level) declined with distance from restored habitat and forest edge. Distance from restored habitat explained more of the variance in bee diversity than distance from forest. Likewise, the richness of butterflies declined with distance from the restored habitat; however, butterfly diversity was unchanged with distance from forest edge.

Cusser and Goodell (2014) studied plots of native meadow on reclaimed surface mineland at The Wilds that ranged in floral diversity and isolation from remnant woodland habitat (Figure 6.3). They found pollinator diversity declined with distance to the remnant habitat, and lower pollinator diversity in low floral diversity plots distant to the remnant stands was associated with lower network stability. High floral diversity plots, however, apparently compensated for losses in pollinator diversity caused by distance from remnant woodlands by attracting

Figure 6.3 (**See color insert.**) A large-scale ecological research project focused on understanding how native pollinators interact with a landscape that has been disturbed from mining and now is recovering through reclamation and restoration efforts. This work illustrates the stability and robustness of plant–pollinator interactions that can be achieved by introducing diverse prairie and meadow plants, even in the face of isolation from remnant habitat. (Photo credit: Joe Clark.)

generalist pollinators. Generalist pollinators increased network connectedness and plant-niche overlap, and consequently, increased network robustness independent of isolation. Therefore, the chance of extinction events, even in isolated plots, was decreased. This work illustrated the stability and robustness of plant–pollinator interactions that can be achieved by introducing a diverse mixed prairie and meadow plants, even in the face of isolation from remnant habitat.

Reptiles and Amphibians

Results from research at The Wilds show the importance of reclaimed minelands to biological diversity for a number of taxa. Reptiles and amphibians are among the most threatened vertebrate groups in the world (Gibbons et al. 2000; Lannoo 2005). Habitat loss has been often cited as a major reason for the declines (Gibbons et al. 2000). Thus, habitat restoration projects, including artificial habitats created on post-mined lands, can be valuable to amphibian and reptile diversity. In Indiana, Lannoo et al. (2009) examined amphibian and reptile species on a large tract of mineland restored to prairie and found 13 species of amphibians (nine frogs and four salamanders) and 19 species of reptiles (one lizard, five turtles, and 13 snakes). This included two state-endangered species and three state species of special concern. Nearly half could be considered obligate grassland species or were common on the edges of grasslands. Lannoo et al. (2009) concluded that carefully implemented prairie restoration on minelands could support roughly the same degree of amphibian richness as native prairie restorations.

Grassland Birds

Grassland birds have been the focus of numerous conservation efforts in the face of habitat loss and increasing nest predation. Grasslands on SMCRA-reclaimed mines, although exotic, are refugia for many grassland bird species, including bird species that are endangered or declining. Similarly, Scott et al. (2002) concluded that obligate grassland birds benefited from the present dominance of non-native grasses on reclaimed mines.

Ingold (2002) documented the presence of 10 grassland and shrub-nesting passerine species on reclaimed surface minelands at The Wilds. The most abundant species were grasshopper sparrows (*Ammodramus savannarum*), Henslow's sparrows (*Ammodramus henslowii*), eastern meadowlarks (*Sturnella magna*), and red-winged blackbirds (*Agelaius phoeniceus*). Grasshopper, Henslow's, and savannah (*Passerculus sandwichensis*) sparrows had the greatest reproductive success. The average rate of nest predation for all species was 39 percent (Ingold 2002). Ingold and Dooley (2013) revisited their study sites to compare reproductive success of ground-nesting and above-ground nesting passerine species, determine nest predation rates of species, and determine if there were differences in predation rates between ground- and shrub-nesting species. They found 18 species, although unlike 10 years prior, they did not find savannah sparrow, Henslow's sparrow, or short-eared owl (*Asio flammeus*) nests. However, they did detect singing Henslow's (Figure 6.4) and savannah sparrows near the surveyed site. These findings suggest that the site continued to provide adequate nesting habitat for most of the obligate grassland species. Ground-nesting species had higher nest success than above-ground and shrub-nesting species. Grasshopper sparrows, eastern meadowlarks, and bobolinks (*Dolichonyx oryzivorus*) had the greatest reproductive success rates, which were comparable to those reported by Ingold in 2002.

Figure 6.4 A male Henslow's Sparrow (*Ammodramus savannarum*) vocalizes nesting territory in the exotic grasslands of The Wilds. (Photo credit: David Cree.)

Predation rates were higher in the above-ground and shrub-nesting (44 percent) versus ground- and near-ground-nesting species (27 percent). Six shrub-nesting species found in this study were not found a decade prior. Although the results suggested the site continues to provide suitable nesting habitat for obligate grassland species, the increasing encroachment of autumn olive was likely attracting more shrub-nesting bird species.

Other extrinsic factors may influence grassland habitat structure (e.g., mowing, hay-cropping, burning, and grazing), thereby affecting reproductive success and/or nest-site fidelity on reclaimed surface mines. Numerous studies have shown that mowing and hay-cropping, particularly between May and August, adversely affect reproductive success of several grassland bird species in the Midwest and northeastern United States. Ingold et al. (2010) addressed the effects of mowing on grassland birds at The Wilds. Mowing did not appear to negatively impact return rates, the act of returning to a previous breeding site, for grasshopper and savannah sparrows, although it may have negatively affected return rates for bobolinks. There are additional management strategies that might enhance habitat quality for grassland birds, including providing perch sites in open areas and increasing floral diversity to support more insects, an important food source for grassland birds, particularly during the nesting season. At The Wilds, the best management strategy for grassland birds appeared to be a varied one because differences in microhabitat improve overall bird diversity. A combination of mowing, burning, and little to no disturbance created a mosaic of grassland habitats to the benefit of numerous species. For example, Henslow's sparrows are not positively associated with disturbed areas and need large, dense thickets of grass and space. Savannah and grasshopper sparrows, however, do not respond negatively to mowing and seem to prefer some disturbance. Bobolinks seem to respond positively to properly-timed burning. The most important factor is having a large expanse of contiguous grassland, regardless of whether the grassland is native or exotic (Ingold et al. 2010).

Although shrubs provide nesting habitats for shrub-nesting species (DeVault et al. 2002; Galligan et al. 2006), several studies have demonstrated that woody encroachment negatively affects grassland-associated bird species (Grant et al. 2004; Graves et al. 2010; Renfrew et al. 2005). Woody vegetation may attract mammalian nest predators, such as raccoons (*Procyon lotor*) and Virginia opossums (*Didelphis virginiana*; Winter et al. 2000).

Based on their work at The Wilds, Ingold and Dooley (2013) stated that if the goal of management on reclaimed mines is healthy populations of grassland-obligate birds, then the encroachment of woody growth should be carefully evaluated. Graves et al. (2010) also reported a negative association between woody vegetation and daily nest survival in grasshopper and Henslow's sparrows on reclaimed minesites in Ohio on Wildlife Management Areas near The Wilds, and suggested efforts to remove woody vegetation. Methods have been developed using The Wilds as a test site for control of autumn olive (Byrd and Cavender 2010) as well as tree of heaven (*Ailanthus altissima*) (Lewis and McCarthy 2006; Peugh et al. 2013).

Predators

Data collected through camera trapping across The Wilds property in 2015 by Driscoll et al. (2017) confirms that the reclaimed minelands at The Wilds support a diverse biological community of birds and mammals. Eighteen mammals and 23 avian species were captured by the camera traps. Of particular interest are insights into the predator community because of the community's association with ecological integrity and biodiversity conservation (Sergio et al. 2006).

As many of the world's top predators are disappearing worldwide, there is growing evidence that such species are critical to ecosystem function, exerting control over smaller predators, prey, and plants (Shurin et al. 2002). Although minelands, such as The Wilds, are not pristine in nature, there is growing evidence that they are providing critical habitat for predators, such as raptors and medium to large mammalian carnivores.

Ingold (2010) surveyed The Wilds from early January through mid-April, 2009 to assess the relative abundance of migratory and winter resident raptors, and to examine potential associations of raptors on reclaimed minelands. He observed nine raptor species (382 sightings), including red-tail hawk (*Buteo jamaicensis*), rough-legged hawk (*Buteo lagopus*), American kestrel (*Falco sparverius*), northern harrier (*Circus cyaneus*), osprey (*Pandion haliaetus*), red-shouldered hawk (*Buteo lineatus*), sharp-shinned hawk (*Accipiter striatus*), short-eared owl (*Asio flammeus*), and golden eagle (*Aquila chrysaetos*). Eastern screech owl (*Magascops asio*) was also observed by Driscoll et al. (2017) between February and July of 2015. These observations indicated a variety of habitat types were utilized. Most notable were the moderately high densities of obligate grassland raptors observed (e.g., rough-legged hawks, short-eared owls, and northern harriers), providing evidence that reclaimed surface mineland is providing valuable habitat for these species, including several that are in decline (Sauer et al. 2008, ODNR 2017). As with grassland bird habitat, Ingold states that if reclaimed surface mines are to maintain healthy populations of winter raptors, it will be important to control the encroachment of woody vegetation.

Coyotes (*Canis latrans*) are more recent additions to the mammalian predator community of the Eastern United States. Today they occur in the majority of the continental United States, Alaska, Canada, Mexico and Central America, and were first observed in the region near The Wilds in 1949 (Weeks et al. 1990). Coyotes were recorded in 79 out of 1314 camera traps at The Wilds from a variety of habitat types on the reclaimed surface mined lands, especially along edges where there was no snow cover (Driscoll et al. 2017). They were often captured on camera in association with turkey vultures (*Cathartes aura*), Virginia opossums, and the American crow (*Corvus brachyrhynchos*), which are considered scavenger species. The authors concluded that coyotes were highly adaptive species that have the ability to inhabit most reclaimed habitats in regions with a viable food source. Other predatory mammalian species captured on the camera traps included the long-tailed weasel (*Mustela frenata*), American mink (*Neovison vison*), red fox (*Vulpes vulpes*), and gray fox (*Urocyon cinereoargenteus*), indicating the diversity of the predator community inhabiting this previously disturbed ecosystem at The Wilds.

Bobcats (*Lynx rufus*) were found throughout Ohio in early settlement times. However, they were extirpated from Ohio during the mid-1800s, and only recently began to recover. An important interaction exists between landscape features of former surface-mined areas and bobcat recovery. Research through the Ohio Division of Wildlife began in 2008 with the primary objective to determine the distribution and relative abundance of bobcats in Ohio. Cameras and scent pads were used to survey bobcats at 12 randomly selected sites in southeastern Ohio, the area that contains the majority of potential bobcat habitat (Figure 6.5). The Wilds was one of only four of these sites with bobcat detections during the initial study. Detection rates were positively correlated with verified sightings within a 5-km radius of the sites. Consequently,

Stealth Cam 06-21-2009 20:16:38

Figure 6.5 A bobcat (*Lynx rufus*) image from camera trapping studies at The Wilds. (Photo credit: Suzanne Prange.)

verified sightings were used as a range-wide index to bobcat distribution and relative abundance. Verified sightings revealed two largely spatially distinct subpopulations within southeastern Ohio. Although initial re-establishment apparently occurred nearly simultaneously within the two areas, the eastern subpopulation increased more rapidly. Furthermore, microsatellite DNA data revealed that the subpopulations were genetically distinct, although within-population genetic variation was high, suggesting limited inbreeding (Anderson et al. 2015). Based on the genetic data, the eastern subpopulation is self-sustaining, whereas the southern subpopulation is maintained through immigration. There are no obvious physical barriers (e.g., waterways, major roadways, human development) to account for the separation of the subpopulations. Coarse-grained habitat variables (e.g., percentage of major habitat types, such as forest or open fields) do not differ between areas used by bobcats versus those unused, and bobcat range across Ohio is positively associated with low densities of human population. However, there was significantly more reclaimed surface mineland in the eastern than southern population area (ODNR, Division of Mineral Resources 2012).

Interestingly, one of the bobcat's primary prey species, the meadow vole (*Microtus pennsylvanicus*), also appears to do well on post-SMCRA reclamation sites. Dooley and Murray (2006) found that there was no correlation between the time since reclamation and the density of meadow voles at three study sites at The Wilds. Conversely, the youngest site exhibited very high vole densities. The three sites were reclaimed in 1969, 1974, and 1984. Thus, only the latter was a post-SMCRA reclamation site. After 15 years of recovery, the authors stated that habitat quality for the meadow vole was much higher than at the older sites. Furthermore, they postulated that the arrested state of succession at this and similar sites might actually benefit the meadow vole, which is a species that is typically displaced as succession to woody vegetation occurs. Benefits to meadow voles may equate to benefits to bobcats and other predators. Rose and Prange (2015) examined bobcat diets in Ohio and found that voles were only consumed by the more-successful eastern population.

Driscoll et al.'s (2017) study of additional monitoring by 10 camera traps randomly deployed across The Wilds' property during a six-month period in 2015 also found interesting and consistent results for bobcats. Bobcats were photographed at seven out of the 10 camera-trap locations and were observed during every month of the study from January to June, although more often in the snow cover and in association where white-tailed deer (*Odocoileus virginianus*), wild turkey (*Meleagris gallopavo*), and the eastern cottontail (*Sylvilagus floridanus*) were also photographed. All of these species are major prey species of bobcats in Ohio (Rose and Prange, 2015).

Bobcats select woodland interspersed with openings, which create cover and a wider prey base, which is consistent with the habitat variables of The Wilds. Black bears (*Ursus americanus*) have a history in Ohio nearly identical to that of bobcats, although their recovery has been more protracted. Bears, like bobcats, select a mixture of woodlands and openings. In this case, the variety of habitats in close juxtaposition helps to ensure year-round food sources. Although it is too early in the recovery of black bears in Ohio to determine whether minelands will be disproportionately occupied, bears are expected to thrive on reclaimed mineland, particularly large areas such as The Wilds.

Summary

Previously mined lands are often dismissed as areas of potential conservation value due to their history of disturbance. However, as demonstrated from observations summarized in this chapter, there is great potential for reclaimed mines to be important areas for the conservation of many trophic levels of biological diversity. This is especially true if concerted management and restoration efforts are implemented, especially when sites are severely disturbed (Table 6.1). Although research should continue to

Table 6.1 Flora and Fauna associated with reclaimed mineland covering multiple trophic levels at The Wilds, management implications, and recommendations

Faunal Group	Habitat Association and Recovery Status	Management Implications and Recommendation	Citation
Cellular Slime Molds	High species richness in remnant forests	Protect forest remnants and extend boundaries through restoration	Cavender and Cavender 2013
Arbuscular Mycorrhizal Fungi	Presence in cool-season exotic grasslands	AMF suitable for native prairie colonization	Thorne et al. 2013
Gastropods	High species richness with both woodland and grassland species present	Maintain a mosaic of habitat for continued recovery; reinforces the need for patches of forest to be incorporated in mining plans	Watters et al. 2005
Butterflies	Moderate species richness in restored native meadows; enhanced stability in high diversity meadows	Improve floral diversity of grasslands and maintain forest habitat	Cusser and Goodell 2014; Cavender-Bares and Cavender 2011; Swab 2017
Bees	High species richness near forest edges and restored meadows	Improve floral diversity of grasslands and maintain forest habitat	Goodell et al. 2010
Grassland birds	Presence of obligate grassland nesting birds	Reduce invasions of woody shrubs and maintain open grasslands through a variety of management techniques	Ingold 2002; Ingold and Dooley 2013
Raptors	Presence of obligate grassland winter raptor species	Reduce invasion of woody shrubs and maintain open grasslands through a variety of management techniques	Ingold 2010; Driscoll 2017
Mammalian predators (e.g., bobcats, black bears, coyote)	Presence of top predators due to diversity of habitats, and presence of cover and prey	Maintain a mosaic of habitat for continued recovery and edge habitat for prey	Anderson and Prange 2015; Rose and Prange 2015;Driscoll 2017

develop a process for converting minelands to forests, reclamation to a healthy forest with diversity at all layers of the forest remains a challenge. There are multiple pathways of recovery after large-scale anthropogenic disturbance. Conversion to diversified native grasslands offers an alternative way to return these lands to highly functioning ecosystems, particularly in the face of the loss of grasslands in North America.

The Wilds, because of its size, history, and organizational focus on conservation and restoration ecology, provides a unique opportunity as a reference to better understand the relationship of natural and anthropogenic disturbances. The Wilds is an interesting case study for how valuable previously mined land can become for biodiversity. Although the pre-disturbance forested ecosystem has not returned, and native vegetation only succeeds with intervention, the reclaimed grassland provides important ecosystem functions (e.g., pollination, decomposition, predation) that have maintained substantial levels of diversity, or even increased in diversity, over a relatively short time period since disturbance. In fact, The Wilds serves as a refuge for several species that are threatened at the state or federal level.

Several factors are likely responsible for the terrestrial ecosystems of The Wilds serving as successful habitat. First, despite much of the area being grassland, remnants of original hardwood forest remain onsite and there is also the presence of reclaimed forest prior to the 1977 SMRCA implementation, which signifies the importance of legacy conditions. These forest remnants likely serve as refugia that enhance re-colonization of surrounding disturbed areas, as well as improve overall landscape connectivity. Reclaimed surface mines that persist as non-forest cover lead to forest fragmentation and reduced connectively. Thus, for non-forest reclamation, it is important to establish or preserve woodland patches for refugia or to serve as stepping stones, or linear corridors for improved connectivity. Second, the land area of The Wilds also includes a variety of habitat types available across an expansive space with low human land-use intensity. Although this increases the number of ecological niches, it also reduces additional complications of urbanization and suburbanization. Third, there has been a concerted effort to maintain this landscape for conservation and protection of wildlife through ecological management.

There are several lessons learned through the management practices at The Wilds. Exotic grasslands are beneficial to grassland species, including grassland-obligate birds, despite the exotic nature of the grasses. However, it is important to control the influx of exotic shrubs, if conserving biological diversity remains a priority. Chemical, mechanical, and prescribed burning regimes should be implemented to support the open-grassland nature, as it is important for nesting establishment. Ideally, incorporating a more-diverse prairie mixture at the onset of reclamation or during restoration management may have more long-term benefits for many biological trophic levels, such as microbes, insects, birds, and even mammals. By introducing prairie species at the onset of reclamation, it would avoid introducing further disturbance during conversion. These observations have implications for a wider context, as we consider and continue to refine ecosystem-level restoration approaches.

Although there is great potential to reclaim sites to functional grassland ecosystems, observations indicate that recovering forests or reclaimed forests are also vital for providing varied habitat types to support a wide range of wildlife. Results from camera traps showed a variety of animals using the edges of the reclaimed forest that were planted prior to SMCRA law. Efforts that support growth of canopy, therefore, should continue to be researched and trialed.

Mined areas cover a substantial land mass and the reclamation process should continue to evolve to meet the optimum level of ecosystem function possible. The Wilds is an ideal place to study changes from disturbance over time and is also a model for ecosystem-level reclamation of a previously mined landscape. This case study has shown the importance of antecedent conditions and past disturbance legacies and the myriad pathways of ecosystem recovery from disturbances. Both natural and anthropogenic disturbances are highly intertwined, and ecosystem functioning can be negatively or positively influenced through human intervention.

Acknowledgements

The authors are greatly indebted to The Wilds, The Columbus Zoo and Aquarium, Muskingum College, The Ohio Department of Natural Resources, and the many dedicated people over the decades who helped contribute to a better understanding of the flora and fauna at The Wilds and of the ecology of reclaimed minelands. The authors would especially like to thank Jessica Turner-Skoff and Shana Byrd for their very helpful comments on the manuscript as well as those of two anonymous reviewers. Our greatest hope is that the information in this chapter helps improve our understanding and approach to reclamation and restoration of disturbed minelands for the support and restoration of biological diversity.

References

Anderson, C. S., S. Prange, H. L. Gibbs. 2015. Origin and genetic structure of a recovering bobcat (*Lynx rufus*) population. *Canadian Journal of Zoology*, 93(11): 889–899.

Bauman, J. M., and N. Cavender. 2011. The Wilds: center for restoration, conservation, and outreach. *Reclamation Matters: American Society of Mining and Reclamation*. Fall: 19–23.

Burger, J. A. 2011. Sustainable mined land reclamation in the eastern US coal-fields: a case for an ecosystem reclamation approach. *Proceedings in The American Society of Mining and Reclamation Proceedings: Sciences Leading to Success.* Lexington, Kentucky.

Burger, J. A., D. Graves, P. N. Angel, V. M. Davis, and C. E. Zipper. 2005. The for-estry reclamation approach. Appalachian Regional Reforestation Institute, US Office of Surface Mining. Forest Reclamation Advisory No. 2.

Byrd, S., and N. Cavender. 2010. Comparison of mechanical, foliar, and dormant stem control methods on mortality of autumn olive (*Elaeagnus umbellate*); a study on reclaimed surface mine land. *Proceedings of the 2010 Ohio Invasive Plant Research Conference.* Columbus OH.

Casselman, C. N., T. R Fox, J. A. Burger, A. T. Jones, and J. M. Galbraith. 2006. Effects of silvicultural treatments on survival and growth of trees planted on reclaimed mine lands in the Appalachians. *Forest Ecology and Management* 223: 403–414.

Cavagnaro, T. R., F. A. Smith, S. E. Smith, and I. Jakobsen. 2005. Functional diversity in arbuscular mycorrhizas: exploitation of soil patches with dif-ferent phosphate enrichment differs among fungal species. *Plant Cell and Environment* 28: 642–650.

Cavender-Bares, J., and N. Cavender. 2011. Phylogenetic structure of plant commu-nities provides guidelines for restoration. In S. Greipsson, editor, *Restoration Ecology,* 119–129. Boston, MA: Jones and Bartlett Publishers.

Cavender, J. C., and C. Hopka. 1986. Distribution patterns of Ohio soil dictyostelids in relation to physiography. *Mycologia* 78: 825–831.

Cavender, J. C., and E. Vadell. 2006. Cellular slime molds of Ohio. *Ohio Biological Survey,* Columbus, Ohio.

Cavender, J. C., and N. D. Cavender. 2013. Butterfly Woods, The Wilds, an optimal habitat of dictyostelidcellular slime molds in Ohio. *Mycosphere* 4: 282–290.

Cavender, N. and S. Byrd. 2007. Use of Roundup Ready® soybeans to reduce Chinese lespedeza (*Lespedeza cuneata*) competition during establishment of high diversity prairie. *Proceedings of the 2007 Ohio Invasive Plant Research Conference, Ohio Biological Survey.* Columbus, Ohio.

Cavender, N., S. Byrd, C. L. Bechtoldt, and J. M. Bauman. 2014. Vegetation commu-nities of a coal reclamation site in Southeastern Ohio. *Northeastern Naturalist* 21: 31–46.

Chorover, J., R. Kretzschmar, F. Garcia-Pichel, and D. L. Sparks. 2007. Soil biogeo-chemical processes within the critical zone. *Elements* 3: 321–326.

Cusser, S., and K. Goodell. 2014. Using a centrality index to determine the contribution of restored and volunteer plants in the restoration of plant-pollinator mutualisms on a reclaimed strip mine. *Ecological Restoration* 32: 179–188.

DeVault, T. L., P. E. Scott, R. A. Bajema, and S. L. Lima. 2002. Breeding bird com-munities of reclaimed coal-mine grasslands in the American Midwest. *Journal of Field Ornithology* 73: 268–275.

Dixon, K. 2009. Pollination and restoration. *Science* 325: 571–573.

Dooley, J. L, and A. L. Murray. 2006. Population responses of *Microtus pennsylva-nicus* across a chronological sequence of habitat alteration. *Ohio Journal of Science* 106: 93–97.

Driscoll, K., M. Lacey, and J. Greathouse. 2017. Use of camera trapping to determine spatial distribution, habitat use, and environmental factors affecting mesopredators on reclaimed mine lands at The Wilds. *Journal American Society of Mining and Reclamation* 6: 15–33.

Galligan, E. W., T. L. DeVault, and S. L. Lima. 2006. Nestingsuccess of grassland and Savanna birds on reclaimed surface coal mines on the midwestern United States. *Wilson Journal of Ornithology* 118: 537–546.

Gibbons, J. W., D. E. Scott, T. J. Ryan, K. A. Buhlmann, T. D. Tuberville, B. S. Metts, J. L. Greene, T. Mills, Y. Leiden, S. Poppy, and C. T. Winne. 2000. The global decline of reptiles, déjà vu amphibians. *BioScience* 50: 653–666.

Gillespie, I. G., and E. B. Allen. 2006. Effects of soil and mycorrhizae from native and invaded vegetation on a rare California forb. *Applied Soil Ecology* 32: 6–12.

Goodell, K., C. Lin, A. McKinney, S. Byrd, B. Bloetscher, and N. Cavender. 2010. Restored and remnant habitat patches feature differently in the movement of bee and butterfly pollinators into adjacent degraded habitat. *Proceedings of the Ecological Society of America 95th Conference*. Pittsburgh, Pennsylvania.

Gorman, J. M., J. G. Skousen, J. Sencindiver, and P. Ziemkiewicz. 2001. Forest productivity and minesoil development under a white pine plantation versus natural vegetation after 30 years. *Proceedings from the 2001 National Meeting of the American Society for Surface Mining and Reclamation*. Albuquerque, New Mexico.

Gragson, T. L., and P. V. Bolstad. 2006. Land use legacies and the future of southern Appalachia. *Society and Natural Resources* 19: 175–190.

Grant, T. A., E. Madden, and G. B. Berkey. 2004. Tree and shrub invasion in northern mixed-grass prairie: implications for breeding grassland birds. *Wildlife Society Bulletin* 32: 807–818.

Graves, B. M., A. D. Rodewald, and S. D. Hull. 2010. Influence of woody vegetation on grassland birds within reclaimed surface mines. *The Wilson Journal of Ornithology* 122: 646–654.

Hammer, C. A. 1984. Dictyostelids in agricultural soils. M.S. thesis, Ohio University, Athens, Ohio.

Holl, K. 2002. Long-term vegetation recovery on reclaimed coal surface-mines in the eastern USA. *Journal of Applied Ecology* 39: 960–970.

Ingold, D. 2002. Use of a reclaimed strip-mine by grassland nesting birds in east-central Ohio. *The Ohio Journal of Science* 102: 56–62.

Ingold, D. 2010. Abundance and habitat use of winter raptors on a reclaimed surface mine in Southeastern Ohio. *The Ohio Journal of Science* 110: 70–76.

Ingold, D., and J. L. Dooley. 2013. Nesting success of grassland and shrub-nesting birds on The Wilds, an Ohio reclaimed surface mine. *The Ohio Journal of Science* 111: 37–41.

Ingold, D., J. L. Dooley, and N. Cavender. 2010. Nest-site fidelity in grassland birds on mowed vs. unmowed areas on a reclaimed stripmine. *Northeastern Naturalist* 17: 125–134.

Jacobs, D. 2005. American Chestnut as a future resource to enhance reclamation productivity. In *Proceedings of the American Society of Mining and Reclamation 22nd Annual National Conference*. Morgantown, West Virginia.

Kaster, G., and J. P. Vimmerstedt. 1996. Tree planting on strip-mined land. In E. R. Norland, and M. S. Ervin, editors, *Forest Resource Issues in Ohio 1996,*

Legislator's Handbook, 2nd Edition. Columbus, OH: Ohio Society of American Foresters.

Kauffman, G. L. 1986. Effects of organic matter and moisture content on dictyostelids in agricultural soil. M.S. thesis, University, Athens, Ohio.

Kearns, C. A., D. W. Inouye, and N. M. Waser. 1998. Endangered mutualisms: the conservation of plant-pollinator interactions. *Annual Review of Ecology and Systematics* 29: 83–112.

Lannoo, M. J., editor. 2005. *Amphibian Declines: The Conservation Status Of United States Species*. Berkeley, CA: University of California Press.

Lannoo, M. J., V. C. Kinney, J. L. Heemeyer, N. J. Engbrecht, A. L. Gallant, and R. W. Klaver. 2009. Mine spoil prairies expand critical habitat for endangered and threatened amphibian and reptile species. *Diversity* 2009: 118–132.

Levy, M. A. and J. R. Cumming. 2014 Development of soils and communities of plants and arbuscular mycorrhizal fungi on West Virginia surface mines. *Environmental Management* 54: 1153–1162.

Lewis, K., and B. McCarthy. 2006. Tree-of-heaven control using herbicide injection. *Ecological Restoration* 24: 54–56.

Lindberg, T. T., E. S. Bernhardt, R. Bier, A. M. Helton, R. Brittany Merola, A. Vengosh, and R. T. Di Giulio. 2011. Cumulative impacts of mountaintop mining on an Appalachian watershed. *Proceedings of the National Academy of Sciences* 108: 20929–20934.

Mummey, D. L., P. D. Stahl, and J. S. Buyer. 2002. Microbial biomarkers as an indicator of ecosystem recovery following surface mine reclamation. *Applied Soil Ecology* 21: 251–259.

Noss, R. F., E. T. LaRoe, and J. M. Scott. 1995. *Endangered Ecosystems of the United States: A Preliminary Assessment of Loss and Degradation*, Vol. 28 pp. 68–80. Washington, D.C.: US Department of the Interior, National Biological Service.

ODNR, Division of Mineral Resources. 2012. Mines of Ohio. Available at https://gis.ohiodnr.gov/MapViewer/?config=OhioMines. Accessed May 2018.

Ohio Department of Natural Resources (ODNR). 2017. *Ohio's Listed Species: Wildlife that are Considered to be Endangered, Threatened, Species of Concern, Special Interest, Extirpated, or Extinct in Ohio*. Publication 5356(R0917), ODNR.

Olson, D. M., E. Dinerstein, E. D. Wikramanayake, N. D. Burgess, G. V. N. Powell, E. C. Underwood, J. A. D'Amico, I. Itoua, H. E. Strand, J. C. Morrison, C. J. Loucks, T. F. Allnutt, T. H. Ricketts, Y. Kura, J. F. Lamoreux, W. W. Wettengel, P. Hedao, K. R. Kassem. 2001. Terrestrial ecoregions of the world: a new map of life on Earth. *Bioscience* 51(11): 933–938.

Palmer, M. A., E. S. Bernhardt, W. H. Schlesinger, K. N. Eshleman, E. Foufoula-Georgiou, M. S. Hendryx, A. D. Lemly, G. E. Likens, O. L. Loucks, M. E. Power, P. S. White, and P. R. Wilcock. 2010. Mountaintop mining consequences. *Science* 327: 148–149.

Peugh, C. M., J. M. Bauman, and S. M. Byrd. 2013. Case study: restoring remnant hardwood forest impacted by invasive tree-of-heaven (*Ailanthus altissima*). *Journal American Society of Mining and Reclamation* 2: 99–112.

Poncelet, D. M., N. Cavender, T. Cutright, J. Senkdo. 2014. An Assessment of microbial communities associated with surface mining-disturbed overburden. *Environmental Monitoring and Assessment* 186: 1917–1929.

Renfrew, R. B., C. A. Ribic, and J. L. Nack. 2005. Edge avoidance by nesting grassland birds: a futile strategy in a fragmented landscape. *Auk* 122: 618–636.

Ricketts, T. H., E. Dinerstein, D. M. Olson, C. J. Loucks, W. Eichbaum, D. A. DellaSala, K. Kavanagh, P. Hedao, P. Hurley, K. Carney, R. Abell, and S. Walters. 1999. *Terrestrial Ecoregions of North America: A Conservation Assessment.* Washington, D.C.: Island Press.

Rose, C., and S. Prange. 2015. Diet of the recovering Ohio bobcat (*Lynx rufus*) with a consideration of two sub-populations. *American Midland Naturalist* 173: 305–317.

Roulston, T. H., and K. Goodell. 2011. The role of resources and risks in regulating wild bee populations. *Annual Review of Entomology* 56: 293–312.

Sauer, J. R., J. E. Hines, and J. Fallon. 2008. *The North American Breeding Bird Survey, Results and Analysis 1966–2007*, Version 6.2.2008. Laurel, MD: USGS Patuxent Wildlife Research Center.

Scott, P. E., T. L. DeVault, R. A. Bajema, and S. L. Lima. 2002. Grassland vegetation and bird abundances on reclaimed Midwestern coal mines. *Wildlife Society Bulletin* 30: 1006–1014.

Sergio, F., I. Newton, and L. Marchesi. 2006. Ecologically justified charisma: preservation of top predators delivers biodiversity conservation. *Journal of Applied Ecology* 43: 1049–1055.

Shurin, J. B. E. T. Borer, E. W. Seabloom, K. Anderson, C. A. Blanchette, B. Broitman, S. D. Cooper, and B. S. Halpern. 2002. A cross-ecosystem comparison of the strength of trophic cascades. *Ecological Letters* 5: 785–791.

Simmons, J., W. Currie, K. N. Eshleman, K. Kuers, S. Monteleone, T. L. Negley, B. R. Pohlad, and C. L. Thomas. 2008. Forest to reclaimed land-use change leads to altered ecosystem structure and function. *Ecological Applications* 18: 104–118.

Skousen, J. G., C. D. Johnson, and K. Garbutt. 1994. Natural revegetation of 15 abandoned mine land sites in West Virginia. *Journal of Environmental Quality* 23: 1224–1230.

Skousen, J., P. Ziemkiewicz, and C. Venable. 2006. Tree recruitment and growth on 20-year-old unreclaimed surface mined lands in West Virginia. *International Journal of Mining, Reclamation, and the Environment* 20: 142–154.

Stein, B. A., L. S. Kutner, J. S. Adams. eds. 2000. *Precious Heritage: The Status of Biodiversity in the United States.* Cary, NC: Oxford University Press.

Swab, R. M., N. Lorenz, S. Byrd, and R. Dick. 2017. Native vegetation in reclamation: Improving habitat and ecosystem function through using prairie species in mine land reclamation. *Ecological Engineering* 108: 525–536.

Surface Mining Control and Reclamation Act (SMCRA). 2006. Public Law 95–87. Surface Mining Control and Reclamation Act of 1977. US Code, Title 30, Chapter 25. USOSM, Washington, D.C.

Thorne, M., and J. Cardina. 2011. Prairie grass establishment on calcareous reclaimed mine soil. *Journal of Environmental Quality* 40: 1824–1834.

Thorne, M., L. Rhodes, and J. Cardina. 2013. Effectivity of arbuscular mycorrhizal fungi collected from reclaimed mine soil and tallgrass prairie. *Open Journal of Ecology* 2013: 224–233.

Tilman, D., and J. A. Downing. 1994. Biodiversity and stability in grasslands. *Nature* 367: 363–365.

Tilman, D., J. Hill, and C. Lehman. 2006. Carbon-negative biofuels from low-input high-diversity grassland biomass. *Science* 314: 1598–1600.

Torbert, J. L., and J. A. Burger. 2000. Forest land reclamation. In R. I. Barnhisel, R. G. Darmody, and W. L. Daniels, editors, *Reclamation of Drastically Disturbed Lands*. Agronomy Monograph No. 41, pp. 371–378. Madison, WI: Soil Science Society of America.

Turner, J. B. 2015. The root of Sustainability: Investigating the relationship between medicinal plant conservation and surface mining in Appalachia. Ph.D. Dissertation, West Virginia University, Morgantown.

U.S. Department of the Interior (USDI). 1979. Permanent Regulatory Program Implementing Section 501(b) of the Surface Mining Control and Reclamation Act of 1977: Environmental Impact Statement. Washington, D.C.

U.S. Government Accountability Office (US GAO). 2009. Characteristics of mining in Mountainous Areas of Kentucky and West Virginia. GAO-10–21.

U.S. Office of Surface Mining (USOSM). 2010. *Annual Evaluation Reports for States and Tribes*. Washington, D.C.: US Department of the Interior.

Vaness, B. M., and S. D. Wilson. 2007. Impact and management of Crested Wheatgrass (*Agropyron cristatum*) in the northern Great Plains. *Canada Journal of Plant Science* 87: 1023–1028.

Vitousek, P. M. 1990. Biological invasions and ecosystem processes: towards an integration of population biology and ecosystem studies. *Oikos* 57: 7–13.

Watters, G. A., T. Menker, and S. H. O'Dee. 2005. A comparison of terrestrial snail faunas between strip-mined land and relatively undisturbed land in Ohio, USA – an evaluation of recovery potential and changing faunal assemblages. *Biological Conservation* 126: 166–174.

Weeks, J. L., G. M. Tori, and M. C. Shieldcastle. 1990. Coyotes (*Canis latrans*) in Ohio. *Ohio Journal of Science* 90: 142–145.

Williams, J., and P. Diebel. 1996. The economic value of prairie. In F. B. Sampson and F. L. Knopf, editors, *Prairie Conservation: Preserving North America's Most Endangered Ecosystem*, 19–35. Covelo, CA: Island Press.

Winter, M., D. H. Johnson, and J. Faaborg. 2000. Evidence for edge effects on multiple levels in tallgrass prairie. *Condor* 102: 256–266.

Yarnell, S. L. 1998. The Southern Appalachians: a history of the landscape. General Technical Report SRS-18. Ashville, North Carolina.

Zipper, C. E., J. A. Burger, J. G. Skousen, P. N. Angel, C. D. Barton, V. Davis, and J. A. Franklin. 2011. Restoring forests and associated ecosystem services on Appalachian coal surface mines. *Environmental Management* 47: 751–765.

chapter seven

Toward a Theory of Connectivity among Depressional Wetlands of the Great Plains

Resiliency to Natural and Anthropogenic Disturbance within a Wetland Network

Gene Albanese and David Haukos

Introduction

Disturbance, across spatial and temporal scales, is a critical component of the ecological function of wetland systems. Natural disturbance can be regular and predictable, such as tidal fluxes for coastal systems; sporadic and seemingly random, such as hurricanes or storm surges; or periodic but indiscriminant in time and space, such as drought (Mitsch and Gosselink 2007). Natural disturbance of wetlands typically results in a temporary change in ecological state. Most common are changes in water levels or between wet and dry ecological states. The ecological capacity of wetlands to support relatively high numbers of biota and biodiversity is directly related to the effects of natural disturbance (e.g., drought, inundation, natural herbivory). Disturbance causes short-term changes in wetland structure and function that creates a cumulative number of niches greater than would be possible for a wetland in a constant ecological state (Batzer and Baldwin 2012). Indeed, the ecological function of many wetland ecosystems depends on the perpetual existence of one or more disturbance factors. Ecological processes of wetlands rooted in nutrient cycling and energy flow (i.e., productivity) require facilitation by disturbance events (Mitsch and Gosselink 2007). Even the loss of unpredictable and periodic disturbance impairs wetland systems by diminishing functional relationships across complex ecological interactions among biotic and abiotic factors forming the wetland system (e.g., Euliss et al. 2004).

Holling (2010: 41) defined ecological stability as "the ability of a system to return to an equilibrium state after a temporary disturbance."

This definition implies that a stable or equilibrium ecological state exists in functional, unaltered wetlands and that a disturbance event causes a short-term perturbation from which the wetland will eventually recover. However, for a wetland to be resilient to disturbance, functional relationships among biotic and abiotic factors must persist during and following the disturbance. Further, for persistence of wetlands in a stable ecological state, any changes in state and driving variables must be absorbed or adapted to, or else the stable ecological state will change (Holling 2010). State variables that may change and influence ecological stability in response to wetland disturbance include hydroperiod (i.e., the duration of inundation or wet state), vegetation associations (e.g., co-occurring plant species in a community), and physical attributes (e.g., depth, volume; Tsai et al. 2007; Smith et al. 2012). Hydrology and associated hydrologic budgets are primary driving variables for wetlands, whereby large perturbations to these variables will cause sudden and permanent changes in ecological states.

Anthropogenic disturbance (e.g., cultivation, hydrological diversion) typically reduces stability and resilience of wetlands by altering relationships among biotic and abiotic factors (e.g., Rehage and Trexler 2006; Wei et al. 2013). Such alterations result in deviations from expected functional relationships. These deviations cause shifts in ecosystem drivers that result in changes to the dominant ecological state and frequency of alternative states (Davis et al. 2013). Identification and classification of disturbance events or conditions as natural or anthropogenic are crucial to understanding the wetland system and developing conservation strategies for particular wetland types. To conserve a functional, ecologically intact wetland, natural disturbance and associated factors of resilience should be retained while minimizing the effect of anthropogenic disturbance that operate to disrupt the stasis of the natural disturbance regimes.

Stability and resilience of wetlands are typically considered on a site or individual-wetland scale (e.g., Frieswyk and Zedler 2006), where measures of functional assessment are defined and restricted to the individual wetland. For example, the Hydrogeomorphic (HGM) Approach to wetland assessment developed to measure change in wetland condition requires a comparison of the ecological function of altered wetlands to unaltered wetlands (Davis et al. 2013). Unfortunately, in the regulatory and loss mitigation process, HGM and other assessment methods can only measure and place values on ecological functions for individual wetlands *in situ* within the defined boundaries of the wetland. Such functional assessment requires direct comparison to a reference or relatively functionally intact wetland of the same type. There is no consideration of the function or ecological value of the wetland for relationships with surrounding landscapes and other individual wetlands, which weakens potential conservation and regulatory benefits.

The concepts of stability, resilience, and ecological function of individual wetlands ignore the value of the wetland within a much larger system formed by the conglomeration of individual wetlands of similar type interacting across time and space. Ecological functions of individual wetlands are just components at small spatial scales within a relatively larger wetland system formed by spatial or topological relationships among wetlands across landscapes. It is at the scale of the system where the true contribution of ecological goods and services by wetlands should be measured and evaluated. System scale is typically user-defined based on the ecological process or species of interest. Therefore, the ecological function of an individual wetland should include a measure of its relative contribution to a wetland system in addition to site-specific provision of ecological goods and services. For example, given that water storage is considered a wetland ecological service, it is the cumulative water storage volume and capacity of a wetland system that provides the value of this ecological service, not the water storage capacity of a single wetland within the system.

Hydrologically connected wetlands (e.g., coastal marsh) are an obvious representative of a functional wetland system. However, cumulative value of hydrologically isolated wetlands are rarely considered an ecological function. Historically, there were hundreds of thousands of isolated, freshwater depressional wetlands across the Great Plains of North America. The two dominant depressional wetland systems in this region are the prairie potholes of the glaciated northern Great Plains and playas of the semi-arid western Great Plains, generally known as the High Plains (Galatowitsch 2012; Smith et al. 2012). Individually, these wetlands are typically small, hydrologically driven by run-off from precipitation events, and subject to extensive anthropogenic disturbance (e.g., Johnson et al. 2012; McCauley et al. 2015; see below). Individual wetlands experience annual and inter-annual variation in inundation patterns and hydroperiods due to natural disturbances caused by short- and long-term drought and spatially discrete extreme precipitation events that are essential to the ecological function of these wetland systems (Euliss et al. 1999; Johnson et al. 2004; Smith et al. 2012). These natural disturbances create a distribution of prairie potholes and playas across landscapes that vary in space and time in spatial configuration and variation in ecological state and hydrological condition to form a collective system supporting biodiversity and the aggregate total of ecological goods and services (Jenkins et al. 2010). However, the effect of the loss of individual wetlands relative to the ability of the system to support biodiversity and provide ecological goods and services is unknown.

There is a distinct ecological difference between prairie potholes and playa wetlands that must be considered when assessing disturbance response and resiliency of these wetland networks. The ecological stable state of most functional prairie potholes is predictable inundation; even potholes classified as temporary or seasonal may have predictable periods of inundation on an annual basis and provide reliable ecological conditions for wetland biota (Mitsch and Gosselink 2007). The natural disturbance for many prairie potholes is a longer-than-usual dry stage resulting from gradual drying of the wetland due to long-term drought (Euliss et al. 2004). At temporal scales of decades or longer, many prairie potholes are predominantly inundated with short periods of drying that are critical to the ecological function of the wetland via oxidation of benthic zones (Murkin 1989; Euliss et al. 1999). This dry state constitutes a short-term disturbance within these wetlands and is a minor disruption to the role of a wetland to this system.

Conversely, inundation by localized, intense precipitation events is the disturbance mechanism for playa wetlands for which the ecological stable state is dry (Smith et al. 2012). The ecological state change from dry to wet is sudden (<24 hours), and rapidly changes the ecological function of the playa. A playa wetland that lacks additional water inputs immediately begins drying, moving it toward a return to the stable dry state. Inundation of individual playa depressions is unpredictable, as the dry state may last for years or decades (Johnson et al. 2011). However, it is these localized precipitation events at broad spatial scales that have a tremendous effect on the connectivity and resiliency of the playa system, because the spatial distribution and number of inundated playa wetlands across the landscape vary on an inter-annual or shorter basis.

Prairie potholes and playas occur within a landscape matrix that has been greatly altered by agricultural activities (e.g., cultivation), and this context has diminished the ecological function of these wetlands in all ecological states (Smith 2003; Dahl 2011; Sylvester et al. 2013). These wetlands are subject to considerable anthropogenic disturbance, due in part to their varying ecological states (i.e., frequent dry periods) and isolation that allows for direct physical alteration of individual wetlands (i.e., filling, cultivation), modification of hydrology (e.g., drainage, redirection of overland flow), and installation of alien features (e.g., roads, powerlines, culverts, irrigation equipment, and other structures; Tiner 2003; Johnson et al. 2012). The anthropogenic disturbance of greatest concern is the filling of wetlands with accumulated sediment transported from adjacent cultivated watersheds via water and wind erosion (Gleason and Euliss 1998; Luo et al. 1999; Smith et al. 2011; Tangen and Gleason 2008; Burris and Skagen 2012). Wetlands filled with sediment have minimal ecological function primarily due to the extensive alteration of hydrology and increased anthropogenic disturbance (Smith et al. 2011). Thus, these individual wetlands contribute little to the system and are effectively lost to the landscape topology (Johnson et al. 2012).

The stable inundated state of prairie potholes has been recognized as ecologically valuable, providing numerous ecological goods and services (e.g., Leitch and Hovde 1996; Gleason et al. 2008, 2011). Tens of millions of dollars have been expended to conserve prairie potholes based on this ecological condition through a variety of initiatives, partnerships, and conservation plans (Doherty et al. 2013). Although general ecological goods and services have been reported for playas (Smith et al. 2011), there has not been a concurrent estimation of their cumulative ecological or economic value similar to the pothole system. Although numerous approaches exist to assess the ecological condition of individual prairie potholes and playas (e.g., Guntenspergen et al. 2002; Johnson 2011), none address the ecological contribution of the wetland to the broad-scale system of prairie potholes or playas.

Conservation efforts for playas using traditional approaches (e.g., opportunistic volunteer conservation, financial incentives) for wetlands have failed (Smith et al. 2012). Therefore, conservation efforts directed at playa wetlands will require a paradigm change from the typical approach to preserving ecological function of wetlands. Prioritizing management actions and developing conservation strategies over broad spatial scales for playa wetlands has been difficult because it is challenging to assess the ecological value of a dry playa depression that is rarely inundated. Following conventional wetland conservation strategies requires prioritization of actions based on the ecological value during a wet state. Overcoming the incorrect assumption that dry playas have no or little ecological value is the primary impediment to conservation of playas.

Both prairie pothole and playa wetland systems have been nationally and internationally recognized as important ecosystems for conservation and human well-being (Millennium Ecosystem Assessment Board 2005; North American Waterfowl Management Plan Committee 2012). Initiatives such as the North American Waterfowl Management Plan, Landscape Conservation Cooperatives, and numerous other groups have developed conservation plans and strategies for prairie potholes and playas, and their associated biota. However, all conservation efforts of isolated, inland wetlands are primarily focused on individual wetlands with discrete boundaries, with the relative ecological value and prioritization for conservation through management or restoration based on site-specific characteristics without regard to the relative ecological value of each wetland to the larger landscape-level system of interacting wetlands. There is also a lack of a comprehensive regulatory framework for the conservation of individual wetlands that considers the natural varying ecological state and contribution of individual wetlands to a system (Haukos and Smith 2003; van der Valk and Pederson 2003). Given that cumulative provision of ecological goods and services is predicated on a system of functional wetlands, ignoring the relative contribution of individual wetlands to the system may result in missed conservation opportunities that could have the greatest ecological benefits for all biota dependent on these systems. Our objectives were to (1) demonstrate the underlying concepts for assessing isolated wetlands as a system of interdependent components responding to disturbance at various temporal and spatial scales, and (2) develop a framework for evaluating natural and anthropogenic disturbance on the ecological value of individual, freshwater, isolated, depressional wetlands to a landscape-scale system. We will use the playa wetland system of the Southern High Plains to illustrate these concepts (Smith et al. 2012).

Playa Wetlands as a System

The historical playa wetland assemblage is estimated to have included approximately 50,000 to 80,000 densely clustered wetlands in a portion of the western Great Plains that encompasses parts of Colorado, Kansas, Nebraska, New Mexico, Oklahoma, and Texas (Guthery and Bryant 1982, Albanese and Haukos 2017). Playas are the predominant surface hydrological feature in the southern portion of this region (Smith 2003). Playa depressions are geographically isolated and have a relatively simple physical structure that is typified as small ($\bar{x}=3.2$ ha, 52 percent <1 ha), shallow (i.e., <1 m), and circular. Playa depressions depend solely on precipitation to inundate and thus have no other natural hydrologic input (e.g., groundwater).

In a playa depression, disturbance in the form of periodic inundation is infrequent and hydroperiods are short-term and unstable (Cooke et al. 2005; Tsai et al. 2007; Johnson et al. 2011). Episodic shifts between wet and dry ecological states are normal, but unpredictable in time and space. During the dominant dry state of many playa depressions, there is a relatively indistinct boundary, beyond soil types, between the depression and surrounding upland matrix. Playa depressions fill and wetland islands temporally materialize on the landscape when a region within this system receives sufficient moisture through periodic and seasonally intensified precipitation patterns. Once inundated, these wetland islands develop abrupt boundaries that transition immediately to upland without gradation. These wet-state playas become the principal habitat for the wealth of wetland-dependent species that breed, migrate through, and winter throughout the Southern Great Plains (Haukos and Smith 1994; Smith et al. 2012).

The distribution of playa wetlands, the biotic communities that use them, and the ecological space in which playa wetland-dependent organisms can survive and reproduce is constrained by a legacy of broad-scale climatic and geomorphologic processes. These processes produced the physical relief, patchy spatial distribution, and dense, clustered configuration of playa depressions across the Southern Great Plains (Smith 2003). Climate drives variable, broad-scale precipitation and temperature patterns. These processes collectively moderate the occurrence, frequency, and turnover rate of wet-state playa depressions across the system while seasonal, often extreme, weather events inundate and desiccate localized regions of playa depressions within the system (Haukos and Smith 1993; Johnson et al. 2011). Thus, the ecological state and attributes of playa depressions vary continuously across this system because these processes are largely unpredictable and operate at different rates across multiple spatiotemporal scales. This creates a hierarchal structure of patchiness within and among playa depressions. In response, populations of playa wetland-dependent species have a pronounced spatial structure at any point in time. However, the occurrence, connectivity, and composition of these populations within and among playa depressions are stochastic in space and time because of the continuous transient dynamics in abiotic and biotic processes that constrain and characterize this system.

Playa Systems Form Networks

The application of network theory and analysis to real-world networks has provided unique information and predictions of pattern/process relationships in many scientific fields, including ecology (Jasny et al. 2009; Dale and Fortin 2010). We can represent this dense, large system of thousands of playa depressions and the millions of potential biological connections among depressions as a spatially explicit, topological network model. Weights reflecting differences in the magnitude of the movements of organisms, propagules, and their genes between playa depressions, as a product of different playa wetland attributes (e.g., wet-state depression and wetland habitat quality), can also be assigned to further parameterize these models. Additionally, habitat (e.g., a wet-state playa depression) is similar among playa wetland-dependent species although dispersal biology differs among species. Referencing systems of geographically-isolated wetlands as a topological network provides an inimitable modeling framework to examine the effect of disturbance patterns on dispersal processes of multiple species when accurate wetland distribution data are available because the distribution is a known, fixed constraint.

Integrating an ecological hierarchy conceptual model with network hierarchal analytical techniques is critical to identifying the emerging properties of a system (Allen and Starr 1982). In particular, the scales at which unique system-wide connectivity patterns emerge as the result of different process rate constraints (e.g., the availability of wet-state playa depressions or the dispersal capacity of different species of organisms), the domains over which these patterns remain stable, and the scale thresholds at which these patterns transition can be identified in this manner (Clauset et al. 2004, 2008; Palla et al. 2005). Scale concepts can be used to decompose the playa wetland system into levels of organization because the scale-dependent global connectivity patterns that characterize the underlying physical structure of the system change the dominant dispersal processes under consideration (Gardner et al. 1987; With and Crist 1995; Keitt et al. 1997). Thus, this modeling framework provides empirical validation for the levels of organization used to simplify the system into a hierarchy and the choice of observational scales used for further analysis.

Network-connectivity metrics have been developed to quantify connectivity within different levels of a network hierarchy (Brandes and Erlebach 2005; Rayfield et al. 2011). Network-level metrics are single measures that summarize the connectivity behavior of the entire network, whereas element-level metrics quantify connectivity between discrete habitat patches based on their position and properties (Estrada and Bodin 2008; Rayfield et al. 2011). However, network- and element-level metrics are not independent because the structure of a network at one level in a hierarchy is nested within higher levels and constrained by lower levels. Thus, the relative importance of individual playa depressions to maintaining system-wide connectivity patterns can be assessed by targeting playa depressions for removal from a network based on the value of element-level metrics. The effect of these removals is then determined by examining the loss of network-level connectivity after their removal. Different metrics also conceptually capture alternate aspects of connectivity and, in particular, several ecological studies have highlighted the importance of habitats with greater element-level centrality to maintaining and enhancing connectivity (Urban and Keitt 2001; Bodin and Norberg 2007). Once the subsystem of wetlands that provide the greatest contribution to maintaining the integrity of system-wide connectivity patterns is identified, these wetlands can be efficiently ranked and displayed on a map to guide conservation initiatives. For example, using network analyses, Albanese and Haukos (2017) combined spatial distribution, disturbance frequency (i.e., inundation), and ecological condition to rank and prioritize the contribution of individual playas to the playa wetland system of the Southern High Plains.

An Interconnected Biological Network

Conceptualizing connectivity among playa wetland populations as a multi-scalar process is imperative to furthering our knowledge on how this system's physical structure and dynamics in wetland resources affect the flow of individuals, populations, and genes among playa wetlands and, ultimately, the persistence of these populations throughout the landscape. Although playa wetlands are geographically isolated, the plant and wildlife populations that inhabit and use these wetlands are connected via dispersal processes across the intervening upland matrix. Potential connections emerge among populations when wet-state playa depressions are within the dispersal capacity of a given species. The sheer number, density and clustered spatial configuration of playa depressions across the system facilitate connectivity among populations for species with different dispersal capacities across spatiotemporal scales. These biological connections between spatially discrete playa wetland populations form a broad-scale, interconnected biological network across the Southern Great Plains.

A structural pattern of redundancy and similarity characterizes the physical structure of the playa wetland system and, in turn, maintains both fine- and broad-scale connectivity patterns within this biological network. Although the probability of a given playa depression becoming inundated is relatively low within an annual cycle (Johnson et al. 2011), even under these conditions a redundant, fine-scale pattern of dense, highly connected, localized clusters and long, interconnected paths of wet-state playa depressions typifies connectivity patterns across the extent of this network. These interrelated patterns of connections are critical to facilitating both short- and long-range dispersal processes and they prevent system-wide failure by minimizing the probability of complete loss in function at any point in time. When relatively few wet-state playa depressions are present across the network (i.e., ≤20 percent), potential connections still exist within localized playa wetland clusters for species with limited dispersal capacity (i.e., <5 km). Simultaneously, connections are also present among more distant localized clusters for species that can disperse greater distances (i.e., >5 km; Albanese and Haukos 2017). Furthermore, as more-localized clusters of playa depressions cycle through wet states over broader temporal scales, any overlap in the spatial distribution and extent of wet-state playa depressions facilitates the flow of populations and their genes to more-distant and temporally isolated playa depressions.

Some Implications for Playa Wetland Species

The implications of these connectivity patterns are different for species that depend on playa wetlands for reproduction than for species that use these wetlands temporally during migration or wintering. The playa wetland system is capable of promoting both the relatively rapid exchange of individuals between populations and the slower system-wide flow of populations and genes for resident plant and wildlife species with limited vagility that reproduce in playa wetlands. For example, the playa wetland system supports several resident genera of amphibians, including North American spadefoot toads (Scaphiopodidae) that require wet-state playa depressions for breeding habitat (Smith 2003). These terrestrial anurans are associated with arid and semi-arid habitats, have physiological constraints that limit dispersal capacity, and thus survive the often extreme duration of wet and dry states within playa depressions, with compensatory mechanisms involving adaptation and life-history characteristics. For these species, multiple, spatially structured, local populations among geographically distinct playa wetlands form metapopulations that are reciprocally linked through dispersal (Gray et al. 2004). Fine-scale population dynamics are largely determined by the successful movements of individuals between wet-state playa depressions within localized metapopulations. The potential for these events is limited within annual breeding cycles by the distribution and extent of localized clusters of wet-state playa depressions. However, the redundant spatial structure and overlap in the dynamic pattern of wet-state playa depressions support the development of a network of alternative "stepping-stone" connectivity pathways across the playa wetland system at decadal to millennial scales. The natural disturbance of inundation creates the connectivity across the landscape. In turn, these connectivity pathways facilitate the slow march of individuals, populations, and genes across this system. Furthermore, these broad- and fine-scale connectivity patterns interact to link these dispersal processes while multi-scale pattern redundancy shortens the duration of temporal isolation among populations. In this manner, the playa wetland network accommodates both the rapid, short-range dispersal movements of individuals required for the persistence of localized spadefoot toad metapopulations and the slower, population-level dispersal processes required for gene flow and range-wide shifts across this system.

The efficiency of the playa wetland network is greater for organisms that can disperse further distances (i.e., ≥16 km; Albanese and Haukos 2017). Accordingly, the underlying physical structure and multi-scale connectivity patterns of the playa wetland network also consistently support the system-wide movements of more-vagile and transient species. For example, each year, the playa wetland system provides wetland stopover habitat for >55 shorebird and waterfowl species during their transcontinental migration (Davis and Smith 1998; Smith 2003; Tsai et al. 2012). These species use wetland habitats opportunistically in the Great Plains (Skagen et al. 2008; Webb et al. 2010; Albanese and Davis 2013). Their movement patterns can vary greatly among species within and among seasonal and annual migration periods because, in the Southern Great Plains, capricious weather patterns produce an unstable wetland system with unpredictable resources (Colwell 2010; Albanese et al. 2012). Additionally, the occurrence, abundance, and diversity of migratory shorebirds and waterfowl increase with greater wetland density and variable wetland conditions within localized wetland clusters in this region (Webb et al. 2010; Albanese and Davis 2015). Thus, the playa wetland system accommodates the movements of large numbers of migrating shorebirds and waterfowl each year among dense, localized wetland clusters even when relatively few (<1 percent) wet-state playa depressions are present across this system.

Redundancy, Resilience Thresholds and System Collapse

The inherit complexity of this system suggests that the unaltered playa wetland network had a built-in critical resilience response to the local loss of wet-state playa depressions. Historically, the playa wetland network was capable of maintaining system-wide function when only a small fraction of playa depressions were present as wet-state. Fine-scale patterns of highly connected, localized wet-state playa depressions developed rapidly at this threshold. This phase transition initiated a positive power-law flux (i.e., scale-free) in system-wide efficiency as the spatial scale of connections among wet-state playa depressions increased beyond this threshold when relatively rare events (e.g., widespread simultaneous playa inundation) occurred (Albanese and Haukos 2017). Historically, the broad spatial extent, number, density, and configuration of playa depressions facilitated the emergence of redundant broad- and fine-scale connectivity patterns that, in turn, accommodated the movements of organisms, propagules, populations, and ultimately genes across the system. Network integrity was continuously maintained because the magnitude of these interrelated patterns insulated this system from regional, periodic droughts. Even temporary losses of system-wide function caused by severe periods of widespread drought were likely ameliorated over time by shifts to wetter phases in the broad-scale climatic processes that constrained the occurrence of wet-state playa depressions. The invulnerability of the historical playa wetland assemblage to the loss of wet-state playa depressions is not unique, as pattern self-similarity is often a property of complex, real-world networks that are resilient to damage (Barrat et al. 2008; Newman 2009). Furthermore, investigations into the structural connectivity patterns of other large, unaltered ephemeral wetland networks also support the suggestion that these networks are very robust to the absence of widespread wet-state depressions as the result of drought (Fortuna et al. 2006).

Arguably, the widespread, potentially permanent and progressive losses of playa wetlands have likely triggered the systematic decline in function of this system for wetland populations. It is estimated that approximately 60 percent of the historical playa wetland assemblage has been lost to anthropogenic activities (Johnson et al. 2012). Most that remain are in a reduced ecological condition and nearly 90 percent are vulnerable to sediment infill as a result of predicted land use and climate change by the year 2100 (Johnson et al. 2012; Burris and Skagen 2012). The underlying structure and redundant fine- and broad-scale connectivity patterns that characterized the historical operating range of this wetland system accommodated the emergence of alternative pathways when a subset of playa depressions were dry. In contrast, the complete loss of a playa depression leads to the more-limiting redistribution of viable connectivity pathways available for dispersal processes. This biological network will approach its critical threshold as more playa wetlands are lost. Once the critical threshold in a network is surpassed, network and ecological theory suggests a cascade of subsequent failures that propagate rapid system-wide collapse (Montoya et al. 2006). Given the current estimates of playa wetland loss, we may expect to observe cascading failures, leading to the macroscopic breakdown of this wetland system's functional connectivity patterns. The legacy of playa wetland losses is likely already unfolding, negatively affecting species that depend on playa wetlands for repro-duction, and species that use these wetlands temporally during the non-reproductive portions of their life history (Smith et al. 2012). The effects of decreased functional connectivity portend negative effects on movement strategies, dynamics and persistence of local populations, decreased bio-diversity, and ultimately the redistribution and system-wide extinction of playa wetland-dependent populations.

Although conceptualizing the playa wetland system as an intercon-nected biological network is vital to furthering our understanding of how the interaction between physical structure and wetland resource dynam-ics affects dispersal processes, representing this system as a topologi-cal network provides additional benefits to conservation planning. This analytical framework allows for both the characterization of multiscale connectivity patterns and the resilience of these connectivity patterns to disturbance because both are affected by a network model's topol-ogy (Opdam et al. 2006; Allen and Holling 2010). Albanese and Haukos (2017) demonstrated the potential effects of the complete loss of playa depressions with greater element-level centrality weighted by habitat quality, as a product of current estimates of sediment infill and annual inundation probability, on playa wetland populations with a maximum dispersal capacity of <5 km. When those playa depressions with greater degree centrality were lost, it significantly decreased the magnitude of the formation of localized clusters of wet-state depressions across the sys-tem and doubled the number required for system-wide paths through the network to emerge. System-wide paths failed to emerge whatsoever, even when all of the remaining wet-state playa depressions were avail-able in the network when those depressions with the greatest between-ness centrality values were lost. The complete loss of playa depressions with greater degree and betweenness centrality had an even greater effect, doubling the number of wet-state depressions required for local-ized clusters to form and entirely eliminating the potential for system-wide paths to emerge. It is estimated that the playa wetland system has already lost >30 percent of these high-centrality wetlands, and of those that remain, 80 percent have been more than halfway infilled with sedi-ment (Burris and Skagen 2012). The potential implications of these esti-mates are dire to the persistence of playa wetland-dependent populations if static conservation efforts do not immediately begin to preserve those playa depressions that remain and restore those depressions that have been lost (Albanese and Haukos 2017).

Cooperative efforts by several conservation organizations (e.g., The North American Waterfowl Management Plan, Playa Lakes Joint Venture, Wetland Reserve Program, U.S. Fish and Wildlife Service, Natural Resources Conservation Service, among others) have attempted to conserve playa wetlands within the Southern Great Plains of North American for over 20 years; however, thus far these efforts have been unsuccessful, owing to complex and multifaceted causes (Haukos and Smith 2003; Smith et al. 2012). A fundamental impediment to effective playa wetland conservation has been the failure to accurately conceptualize and integrate the role of spatiotemporal dynamics and scale (i.e., frequency of natural disturbance) when considering, planning, and managing playa wetlands for conservation. We propose that further advances toward the widespread conservation management of playa wetlands will continue to be inhibited unless these limitations are considered. We suggest that successful conservation management of playa wetlands will require a conceptual shift in wetland ecology and conservation. This revised approach should integrate the traditional view of playa wetlands as discrete units with the perspective of this system as an interconnected, wetland biological network. Here, each playa wetland is a component functioning within and contributing to overall system function and persistence. Finally, differentiation of the type of natural disturbance (e.g., wet versus dry stable ecological states) among wetland systems and between natural and anthropogenic disturbance regimes are crucial for development of conservation strategies.

Conclusions

Natural disturbance is critical to the function of depressional, isolated wetlands of the Great Plains. Fluctuating water levels (i.e., prairie potholes) or abrupt changes of ecological states through inundation (i.e., playas) are necessary for the function of individual wetlands. The spatiotemporal variation of ecological states across the landscape form wetland networks that are foundational to cumulative provision of ecological goods and services by isolated wetlands. However, anthropogenic disturbance, primarily in the form of filling through sediment accretion, is causing loss of individual wetlands from the network, which causes cascading negative effects on the integrity of the ecological network. We proposed a novel theoretical model of the playa wetland system as an interconnected ecological network. This model is grounded in the interactions between disturbance and dispersal processes and their constraints across spatiotemporal scales. The analytical framework we suggest is flexible and can provide insights at both the element-specific and system-wide levels. Progression of this framework from the academic to the practical will require greatly improved empirical data on the system-wide distribution and ecological condition of these isolated, freshwater, depressional wetlands in addition to population trends and genetic patterns among populations that depend on these wetlands. If successful, this framework incorporating the role of natural disturbance will help identify the nature of functional connectivity for multiple wetland-dependent species, facilitate predictions of the consequences of anthropogenic and natural disturbances, and provide a means to evaluate the ecological value of individual wetlands to the broader system of wetlands across the Great Plains. Given the current progress of wetland conservation, this framework may have much to contribute towards better decisions about wetland conservation and landscape-level wetland reserve design.

Acknowledgments

This study was funded by National Science Foundation (MacroSystems Biology: Climatic and Anthropogenic Forcing of Wetland Landscape Connectivity in the Great Plains, Award Number 1340646) and U.S. Fish and Wildlife Service, Great Plains Landscape Conservation Cooperative research grants administered through the U.S. Geological Survey Fort Collins Science Center and Kansas Cooperative Fish & Wildlife Research Unit. We gratefully acknowledge additional support provided by the Division of Biology at Kansas State University and Playa Lakes Joint Venture. Comments by S. Skagen, L. Johnson, W. Conway, C. Davis, and E. Beever improved initial drafts of the manuscript. Any use of trade, firm, or product names is for descriptive purposes only and does not imply endorsement by the U.S. Government.

References

Albanese, G., and C. A. Davis. 2013. Broad-scale relationship between shorebirds and landscapes in the Southern Great Plains. *Auk* 130: 88–97.

Albanese, G., and C. A. Davis. 2015. Characteristics within and around stopover wetlands used by migratory shorebirds: is the neighborhood important? *Condor* 117: 328–340.

Albanese, G., C. A. Davis and B. W. Compton. 2012. Spatiotemporal scaling of North American continental interior wetlands: implications for shorebird conservation. *Landscape Ecology* 27: 1465–1479.

Albanese, G., and D. A. Haukos. 2017. A framework for understanding dynamic dispersal processes for multiple species within dense, broad-scale habitat networks: prioritizing wetland conservation in the Great Plains. *Landscape Ecology* 32: 113–130.

Allen, C. R., and C. S. Holling. 2010. Novelty, capacity and resilience. *Ecology and Society* 15(3): 24. www.ecologyandsociety.org/vol.15/iss3/art24/.

Allen, T. F. H., and T. B. Starr. 1982. *Hierarchy: Perspectives for Ecological Complexity.* Chicago, IL: University of Chicago Press.

Barrat, A., M. Barthélemy, and A. Vespignani. 2008. *Dynamical Processes on Complex Networks.* Cambridge, UK: Cambridge University Press.

Batzer, D. P., and A. H. Baldwin. 2012. *Wetland Habitats of North America.* Berkeley, CA: University of California Press.

Brandes, U., and T. Erlebach. 2005. *Network Analysis: Methodological Foundations.* Berlin, Germany: Springer-Verlag.

Bodin, O., and J. Norberg. 2007. A network approach for analyzing spatially structured populations in fragmented landscape. *Landscape Ecology* 22: 31–44.

Burris, L., and S. K. Skagen. 2012. Modeling sediment accumulation in North American playa wetlands in response to climate change, 1940–2100. *Climatic Change* 117: 69–83.

Clauset, A., C. Moore, and M. E. J. Newman. 2008. Hierarchical structure and the prediction of missing links in networks. *Nature* 453: 98–101.

Clauset, A., M. E. J. Newman, and C. Moore. 2004. Finding community structure in very large networks. *Physics Review Letters* 70: 66111–66115.

Cooke, G. D., E. B. Welch, S. A. Peterson, and P. R. Neworth. 2005. *Restoration and Management of Lakes and Reservoirs,* third edition. Baca Raton, FL: Lewis Publishing.

Colwell, M. A. 2010. *Shorebird Ecology, Conservation and Management.* Berkeley, CA: University of California Press.

Dahl, T. E. 2011. *Status and Trends of Wetlands in the Conterminous United States 2004 to 2009.* Washington, D.C.: U.S. Department of the Interior, Fish and Wildlife Service.

Dale, M. R. T., and M. J. Fortin. 2010. From graphs to spatial graphs. *Annual Review of Ecology, Evolution Systematics* 41: 21–38.

Davis, C. A., D. Dvorett, J. R. Bidwell, and M. M. Brinson. 2013. Hydrogeomorphic classification and functional assessment. In J. T. Anderson and C. A. Davis, editors, *Wetland Techniques. Volume 3: Applications and Management,* 29–68. New York: Springer.

Davis, C. A., and L. M. Smith. 1998. Ecology and management of migrant shorebirds in the playa lakes region of Texas. *Wildlife Monographs* 140.

Doherty, K. E., A. J. Ryba, C. L. Stemler, N. D. Niemuth, and W. A. Meeks. 2013. Conservation planning in an era of change: state of the U.S. Prairie Pothole Region. *Wildlife Society Bulletin* 37: 546–563.

Estrada, E., and O. Bodin. 2008. Using network centrality measures to manage landscape connectivity. *Ecological Applications* 18: 1810–1825.

Euliss, N. H., D. M. Mushet, and D. A. Wrubleski. 1999. Wetlands of the Prairie Pothole Region: invertebrate species composition, ecology, and management. In D. P. Batzer, R. B. Rader, and S. A. Wissinger, editors, *Invertebrates in Freshwater Wetlands of North America: Ecology and Management,* 471–514. New York: John Wiley & Sons.

Euliss, N. H., Jr., J. W. LaBaugh, L. H. Fredrickson, D. M. Mushet, M. K. Laubhan, G. A. Swanson, T. C. Winter, D. O. Rosenberry, and R. D. Nelson. 2004. The wetland continuum: a conceptual framework for interpreting biological studies. *Wetlands* 24: 448–458.

Fortuna, M. A., C. Gomez-Rodriguez, and J. Bascompte. 2006. Spatial network structure and amphibian persistence in a stochastic environment. *Proceedings of the Royal Society of London Biology* 273: 1429–1434.

Frieswyk, C. B., and J. B. Zedler. 2006. Do seed banks confer resilience to coastal wetlands invaded by *Typha* x *glauca*? *Canadian Journal of Botany* 84: 1882–1893.

Galatowitsch, S. 2012. Northern Great Plains wetlands. In D. P. Batzer, and A. H. Baldwin, editors, *Wetland Habitats of North America*, 283–298. Berkley, CA: University of California Press.

Gardner, R. H., B. T. Milne, M. J. Turner, and R. V. O'Neill. 1987. Neutral models for the analysis of broad-scale landscape pattern. *Landscape Ecology* 1: 19–28.

Gleason, R. A., and N. H. Euliss, Jr. 1998. Sedimentation of prairie wetlands. *Great Plains Research* 8: 97–112.

Gleason, R. A., M. K. Laubhan, and N. H. Euliss, Jr., editors. 2008. Ecosystem services derived from wetland conservation practices in the United States Prairie Pothole Region with an emphasis on the U.S. Department of Agriculture Conservation Reserve and Wetlands Reserve Programs. Professional Paper 1745. U.S. Geological Survey, Reston, Virginia, USA.

Gleason, R. A., N. H. Euliss, Jr., B. A. Tangen, M. K. Laubhan, and B. A. Browne. 2011. USDA Conservation program and practice effects on wetland ecosystem services in the Prairie Pothole Region. *Ecological Applications* 21: S65–S81.

Gray, M. J., L. M. Smith, and R. I. Leyva. 2004. Influence of agricultural landscape structure on a Southern High Plains, USA amphibian assemblage. *Landscape Ecology* 19: 719–729.

Guntenspergen, G. R., S. A. Peterson, S. G. Leibowitz, and L. M. Cowardin. 2002. Indicators of wetland condition for the Prairie Pothole Region of the United States. *Environmental Monitoring and Assessment* 78: 229–252.

Guthery, F. S., and F. C. Bryant. 1982. Status of playas in the Southern Great Plains. *Wildlife Society Bulletin* 12: 227–234.

Haukos, D. A., and L. M. Smith. 1993. Moist soil management of playa lakes for migrating and wintering ducks. *Wildlife Society Bulletin* 21: 288–298.

Haukos, D. A., and L. M. Smith. 1994. The importance of playa wetlands to biodiversity of the Southern High Plains. *Landscape and Urban Planning* 28: 83–98.

Haukos, D. A., and L. M. Smith. 2003. Past and future impacts of wetland regulations on playas. *Wetlands* 23: 577–589.

Holling, C. S. 2010. Resilience and stability of ecological systems. In L. H. Gunderson, C. R. Allen, and C. S. Holling, editors, *Foundations of Ecological Resilience*, 19–49. Washington, D.C.: Island Press.

Jasny, B. R., L. M. Zahn, and E. Marshall. 2009. Connections. *Science* 325: 405.

Jenkins, W. A., B. C. Murray, R. A. Kramer, and S. P. Faulkner. 2010. Valuing ecosystem services from wetlands restoration in the Mississippi Alluvial Valley. *Ecological Economics* 69: 1051–1061.

Johnson, L. 2011. Current status and function of playa wetlands on the Southern Great Plains. Dissertation, Texas Tech University, Lubbock, Texas, USA.

Johnson, L. A., D. A. Haukos, L. M. Smith, and S. T. McMurry. 2012. Physical loss and modification of Southern Great Plains playas. *Journal of Environmental Management* 112: 275–283.

Johnson, W. C., S. E. Boettcher, K. A. Poiani, and G. R. Guntenspergen. 2004. Influence of weather extremes on the water levels of glaciated prairie wetlands. *Wetlands* 24: 385–398.

Johnson, W. P., M. B. Rice, D. A. Haukos, and P. Thorpe. 2011. Factors influencing the occurrence of inundated playa wetlands during winter on the Texas High Plains. *Wetlands* 31: 1287–1296.

Keitt, T., D. Urban, and B. Milne. 1997. Detecting critical scales in fragmented landscapes. *Conservation Ecology* 1: 4.

Leitch, J. A., and B. Hovde. 1996. Empirical valuation of prairie potholes: five case studies. *Great Plains Research* 6: 25–39.

Luo, H. R., L. M. Smith, D. A. Haukos, and B. L. Allen. 1999. Sources of recently deposited sediments in playa wetlands. *Wetlands* 19: 176–181.

McCauley, L. A., M. J. Anteau, M. Post van der Burg, and M. T. Wiltermuth. 2015. Land use and wetland drainage affect water levels and dynamics of remaining wetlands. *Ecosphere* 6(6): 92 http://dx.doi.org/10.1890/ES14-00494.1.

Millennium Ecosystem Assessment Board. 2005. *Ecosystems and Human Well-being: Wetlands and Water Synthesis*. Washington, D.C.: World Resources Institute.

Mitsch, W. J., and J. G. Gosselink. 2007. *Wetlands*, fourth edition. Hoboken, NJ: John Wiley & Sons.

Montoya, J. M., J. M. Pimm, and R. V. Sole. 2006. Ecological networks and their fragility. *Nature* 442: 259–264.

Murkin, H. R. 1989. The basis for food chains in prairie wetlands. In A. G. van der Valk, editor, *Northern Prairie Wetlands*, 316–339. Ames, IA: Iowa State University Press.

Newman, M. E. J., A. L., Barabási, and D. J. Watts. 2009. *The Structure and Dynamics of Networks*. Princeton, NJ: Princeton University Press.

North American Waterfowl Management Plan Committee. 2012. *North American Waterfowl Management Plan 2012: People Conserving Waterfowl and Wetlands*. Washington, D.C.: U.S. Fish and Wildlife Service, Department of Interior. www.nawmprevision.org/sites/default/files/NAWMP-Plan-EN-may23.pdf.

Opdam, P., E. Steingrover, and S. van Rooij. 2006. Ecological networks a spatial concept for multi-actor planning of sustainable landscapes. *Landscape and Urban Planning* 75: 322–332.

Palla, G., I. Der'enyi, I. Farkas, and T. Vicsek. 2005. Uncovering the overlapping community structure of complex networks in nature and society. *Nature* 435: 814–818.

Rayfield, B., M. J. Fortin, and A. Fall. 2011. Connectivity for conservation: a framework to classify network measures. *Ecology* 92: 847–858.

Rehage, J. S., and J. C. Trexler. 2006. Assessing the nest effect of anthropogenic disturbance on aquatic communities in wetlands: community structure relative to distance from canals. *Hydrobiologia* 569: 359–373.

Skagen, S. K., D. A. Granfors, and C. P. Melcher. 2008. On determining the significance of ephemeral continental wetlands to North American migratory shorebirds. *Auk* 125: 20–29.

Smith, L. M. 2003. *Playas of the Great Plains*. Austin, TX: University of Texas Press.

Smith, L. M., D. A. Haukos, S. T. McMurry, T. LaGrange, and D. Willis. 2011. Ecosystem services provided by playas in the High Plains: potential influences of USDA conservation programs. *Ecological Applications* 21: S82–S92.

Smith, L. M., D. A. Haukos, and S. T. McMurry. 2012. High Plains playas. In D. P. Batzer, and A. H. Baldwin, editors, *Wetland Habitats of North America*, 299–312. Berkeley, CA: University of California Press.

Sylvester, K. M., D. G. Brown, G. D. Deane, and R. N. Kornak. 2013. Land transitions in the American plains: multilevel modeling of drivers of grassland conversion (1956–2006). *Agricultural Ecosystems and the Environment* 168: 7–15.

Tangen, B. A., and R. A. Gleason. 2008. Reduction of sedimentation and nutrient loading. In R. A. Gleason., M. K. Laubhan, and N. H. Euliss, Jr., editors, *Ecosystem Services Derived from Wetland Conservation Practices in the United States Prairie Pothole Region with an Emphasis on the U.S. Department of Agriculture Conservation Reserve and Wetlands Reserve Programs*, 38–44. Professional Paper 1745. Reston, VA: U.S. Geological Survey.

Tiner, R. W., Jr. 2003. Geographically isolated wetlands of the United States. *Wetlands* 23: 494–516.

Tsai, J. S., L. S. Venne, L. M. Smith, S. T. McMurry, and D. A. Haukos. 2012. Influence of local and landscape characteristics on avian richness and density on wet playas in the southern Great Plains. *Wetlands* 32: 605–613.

Tsai, J. S, L. S. Venne, S. T. McMurry, and L. M. Smith. 2007. Influences of land use and wetland characteristics on water loss rates and hydroperiods of playa in the Southern High Plains, USA. *Wetlands* 27: 683–692.

Urban, D., and T. Keitt. 2001. Landscape connectivity: a graph-theoretic perspective. *Ecology* 82: 1205–1218.

Van der Valk, A. G., and R. L. Pederson. 2003. The SWANCC decision and its implication for prairie potholes. *Wetlands* 23: 590–596.

Webb, E. B., L. M. Smith, M. P. Vrtiska, and T. G. LaGrange. 2010. Effects of local and landscape level variables on wetland habitat use during migration through the Rainwater Basin. *Journal of Wildlife Management* 74: 109–119.

Wei, L., T. Rui, W. Jaun, D. Fan, and Y. YuMing. 2013. Effects of anthropogenic disturbance on richness-dependent stability in Napahai plateau wetland. *Chinese Science Bulletin* 58: 4120–4125.

With, K. A., and T. O. Crist. 1995. Critical thresholds in species' responses to landscape structure. *Ecology* 76: 2446–2459.

Index

Page numbers in **bold** refer to tables and those in *italic* refer to figures.

Printed and bound by CPI Group (UK) Ltd, Croydon, CR0 4YY

01/11/2024

01782637-0008